Development with Sustainable Use of Electricity

NATO ASI Series

Advanced Science Institute Series

A Series presenting the results of activities sponsored by the NATO Science Committee, which aims at the dissemination of advanced scientific and technological knowledge, with a view to strengthening links between scientific communities.

The Series is published by an international board of publishers in conjunction with the NATO Scientific Affairs Division

A Life Sciences **B Physics**	Plenum Publishing Corporation London and New York
C Mathematical and Physical Sciences **D Behavioural and Social Sciences** **E Applied Sciences**	Kluwer Academic Publishers Dordrecht, Boston and London
F Computer and Systems Sciences **G Ecological Sciences** **H Cell Biology** **I Global Environment Change**	Springer-Verlag Berlin, Heidelberg, New York, London, Paris and Tokyo

PARTNERSHIP SUB-SERIES

1. Disarmament Technologies	Kluwer Academic Publishers
2. Environment	Springer-Verlag / Kluwer Academic Publishers
3. High Technology	Kluwer Academic Publishers
4. Science and Technology Policy	Kluwer Academic Publishers
5. Computer Networking	Kluwer Academic Publishers

The Partnership Sub-Series incorporates activities undertaken in collaboration with NATO's Cooperation Partners, the countries of the CIS and Central and Eastern Europe, in Priority Areas of concern to those countries.

NATO-PCO-DATA BASE

The electronic index to the NATO ASI Series provides full bibliographical references (with keywords and/or abstracts) to about 50,000 contributions from international scientists published in all sections of the NATO ASI Series. Access to the NATO-PCO-DATA BASE is possible via a CD-ROM "NATO Science and Technology Disk" with user-friendly retrieval software in English, French, and German (©WTV GmbH and DATAWARE Technologies, Inc. 1989). The CD-ROM contains the AGARD Aerospace Database.

The CD-ROM can be ordered through any member of the Board of Publishers or through NATO-PCO, Overijse, Belgium.

3. High Technology – Volume 56

Development with Sustainable Use of Electricity

edited by

Jørgen S. Nørgård

Department of Buildings and Energy,
Technical University of Denmark,
Lyngby, Denmark

Adam Gula

University of Mining and Metallurgy and
The Polish Foundation for Energy Efficiency,
Kraków, Poland

and

Aníbal T. De Almeida

Department of Electronical Engineering,
University of Coimbra,
Coimbra, Portugal

Springer-Science+Business Media, B.V.

Proceedings of the NATO Advanced Research Workshop on
Development with Sustainable Use of Electricity
Lopuszna, Poland
17-21 June 1996

A C.I.P. Catalogue record for this book is available from the Library of Congress.

ISBN 978-94-010-6139-1 ISBN 978-94-011-5092-7 (eBook)
DOI 10.1007/978-94-011-5092-7

Printed on acid-free paper

TABLE OF CONTENTS

PREFACE

In all European countries there is a substantial potential for saving electricity by implementing more efficient end-uses of the electricity. This can significantly reduce the environmental problems, which to a large extent are associated with the generation of electric power. While the saving potentials in Western Europe are in the order of magnitude of half the present electricity consumption, the potentials are even larger in countries in Central and Eastern Europe. Furthermore, since electricity savings are often much cheaper than providing the power, the efficient end-use of electricity can play a vital role in the transition process of the Central and Eastern European countries, because more capital can then be devoted to other essential areas than building and operating power plants, such as education, infrastructure, health, or cleaning the environment.

On this background, a workshop on development with sustainable use of electricity was conducted in June 1996 in Lopuszna, Poland, with most participants coming from the Central and Eastern Europe. The program included lectures, roundtable debates and informal discussions. After some overview lectures which placed the electricity savings in a larger perspective, a number of cases of end-use technologies were presented, including analyses as well as real life experiences. National cases presented the electricity consumption situation and saving policies in Eastern and Western European countries. Barrieres to the required technological and structural changes were discussed as well as policies to overcome those barriers.

The editors as well as the other two members of the Organizing Committee, Jacques Roturier and Arthur Rosenfeld, wish to acknowledge the financial support from the NATO Scientific Affairs Division as the main sponsor, making this Advanced Research Workshop possible. The travel support given by the Commission of the European Communities was also most useful for attracting relevant participants, and is very much acknowledged.

Several individuals have been involved in running the workshop as well as in editing this book with the main lectures of the workshop. We particularly acknowledge the dedicated contributions by Ela Gula, Jaquelin Cochran, Paula Fonseca, Gitte Nellemose, and Anna Levin-Jensen.

The editors hope that the availability of the book will enable a wider audience to have access to the state of the art methodologies and technologies which can integrate the use of electricity in a sustainable development.

The Editors

Jørgen S. Nørgård
Adam Gula
Anibal T. De Almeida

April 1998

SUSTAINABLE USE OF ELECTRICITY

JØRGEN S. NØRGÅRD
Department of Buildings and Energy
Technical Univercity of Denmark
DK-2800 Lyngby, Denmark

1. The Quest for a New Course

In this first contribution to the workshop, I will try to provide a broader background for the topic, *development with sustainable use of electricity*. This topic reflects the necessity for a dramatic change in the course of technological and economical development as compared to the development dominating the last two centuries of industrialization in Europe.

1.1. EARLY WARNINGS

So far the economic development has been dominated by a perception of the world blessed with unlimited options for extracting resources and depositing waste, as indicated in Figure 1, borrowed from Goodland[1]. Today the human economy is often perceived as an omnipotent system which can develop on its own. In the real world, however, the economy is a subsystem of the ecological system and totally dependent on it. High-quality energy and raw materials are extracted from this ecological system and low-grade energy and other waste and pollution are emitted into it. As long as the man made economy was small relative to the ecological system as a whole, the assumption of the unlimited world might have seemed reasonable, although some pioneer economists like M. Stuart Mills and M. J. Keynes were foresighted enough, 150 and 70 years ago respectively, to warn that economic expansion should not and could not go on forever since this would pose a threat to both the quality of human lives and to the environment[2,3,4]. But very few listened to these visionary people, and most economists today are "brought up" with the idea of unlimited natural resources available for free as the foundation on which their education is built. And, as we all know, it is very difficult to change the foundation later on, when the building is finished.

As the economy grew, it occupied a larger and larger part of the ecological system, and today the above decribed view of unlimited resources no longer holds. Already, the present economic activities of the quarter of the world population living in

S. Nørgård et al. (eds.), Development with Sustainable Use of Electricity, 1-18.

the industrialized nations have exceeded the carrying capacity of the environment. Their development is not sustainable, so changes will come, either forced upon us by environmental catastrophes, or as a gentle and well-planned transition.

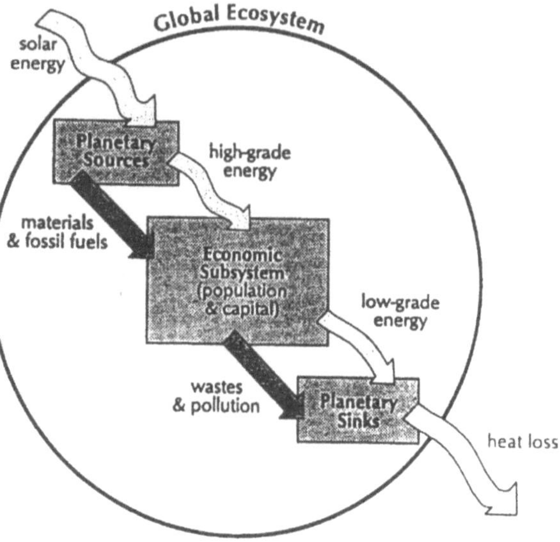

Figure 1.
Qualitative illustration of the fact that the man-made economy is a subsystem of the global ecological system, from which the economy extracts energy and material resources, and into which the economy discharges waste.

Symptoms of the fact that we are approaching or exceeding the limitations are appearing every day at an increasing rate. Ground water is being found polluted, forests and cultivated land are being turned into deserts, ozone layer is being depleted, etc. We will here focus on one global environmental threat, the approaching climatic changes caused by man-made emission of greenhouse gases, especially the CO_2-emission from burning fossil fuels, such as coal, oil and natural gas.

It is important, however, to stress that CO_2-emission is only one out of a number of serious environmental problems, and that these problems are not just accidental incidents, caused by ignorant technological choices. Rather they are symptoms of a philosophy of development, which is basically unsustainable. The CO_2-emission is presently in focus as an indicator of the environmental problems. Also, it just happens that this year it is one hundred years since the Swedish Nobel Prize winner in physics, Arrhenius, warned against the risk of climate changes because of the CO_2-emission from burning fossil fuels[5].

1.2. SUSTAINABLE DEVELOPMENT

The concept of sustainability has for long been used in the language of economics. An economy, in which the income (and consumption) is based on eating up the capital which provides the income, is not sustainable. And this is more or less what we are doing today with our environmental capital. H. Daly, an American economist, has defined a sustainable development with respect to energy and other resources as a development which does not reduce the development options of the future by deteriorating the natural environment[6]. In more operational terms the conditions for a sustainable development can be expressed as follows:

1) Renewable resources (like biomass) should not be used at a faster rate than they regenerate.
2) Pollutants (like CO_2) should not be emitted faster than the environment can absorb or neutralize them.
3) Non-renewable resources (like coal, oil, and gas) should not be used A softer definition of sustainability says that they should not be used at a rate faster than substitutes in the form of energy savings or renewable energy supply systems are built up.

As pointed out, our present use of the environment is violating all the three above rules and is highly unsustainable. As far as CO_2-emission is concerned, the international UN expert panel IPCC has reached the conclusion that on a global scale this emission must be reduced by 65% in the course of a few decades[7]. Even this reduction will most likely not be sufficient to avoid some climatic changes, but it is a compromise target set to keep the risks of the climatic changes at a reasonable level.

1.3. ECOLOGICAL SPACE CONCEPT

A concept "*ecological space*" has been introduced to illustrate the resource use and the pollution that could be allowed per person on earth if we should approach sustainability. The ultimate ecological space for CO_2-emission seems to be around 1.5 tons per capita per year, depending on the population development among other things. The Danish government has adopted the concept of ecological space, despite the fact that for Denmark, with one of the world's highest CO_2-emissions per capita, this will require around 90% reduction in the future. Figure 2 shows the per capita emission for various countries, taken from World Resources[8], plus the maximum permissible level. To reach this level, which is based on an equal right for all human beings to pollute, the industrialized countries are facing an enormous challenge. But fortunately, as illustrated in some contributions to this workshop, it is not at all an impossible task.

The main factors of the problems, and hence also the sources of the solutions, are the development in the following: 1) population, 2) material standard of living, and 3) the technology we apply. As we shall see, the latter option, the technology, holds significant potentials, but we will make a serious mistake by assuming that technological fixes alone can bring us on a path of an environmentally sustainable development, as discussed in a later chapter[9].

CO_2 per capita in selected countries in 1991

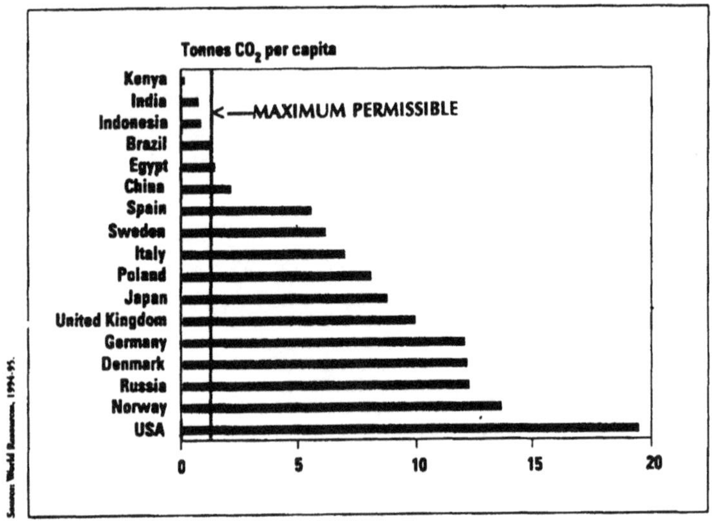

Figure 2.
Annual emission of CO_2 per capita for various countries[8], compared with the maximum emission permissible according to international expert panel IPCC[7], assuming equal right to pollute.

2. Electricity and the Environment

As an energy form electricity is characterized by obvious advantages, mainly by being easily transferable, even over long distances. Also electricity is usable for providing almost any energy service. Furthermore electricity is essential for many electronic systems, which can help *save* electricity and other forms of energy. Hence there is no doubt that electricity plays an important role in providing a sustainable development.

The fact that electricity is needed should not, however, lead to the conclusion that the more electricity consumed the better is the development. The advantages from using electricity must be balanced by a number of disadvantages, which makes the saving and careful use of electricity an essential goal.

2.1. ENVIRONMENTAL PROBLEMS WITH ELECTRICITY

Electricity is often wrongly marketed and perceived as a clean form of energy, mainly because the pollution and other environmental damage occur far from where the electricity is used.

For Western Europe roughly one third of the electricity is produced with coal as the primary energy source and another third from nuclear. Hydropower provides one fifth with the remaining 15% from oil and natural gas. In other words about half of the electricity is made from the CO_2-generating fossil fuels.

If we look at what causes the man-made greenhouse effect with its climatic changes, there are four main sources, namely:

1)	CO_2 from burning fossil fuels	47%
2)	CO_2 from deforestation	15%
3)	CO_2 from cement production	2%
4)	CH_4 from agriculture, waste dumps, etc.	20%
5)	CFCs (Freon, etc)	12%
6)	N_2O	4%

Since typically between one third and half of a country's fossil fuel consumption is used in power plants, the CO_2-emission associated with *the production of electricity will account for around a quarter of the greenhouse gases released.*

With present power production systems electricity ranks among the most polluting and otherwise environmentally harmful forms of energy used. Not only coal, oil and natural gas fired power plants are polluting the air with CO_2, SO_2, NO_x, many heavy metals, etc. Also nuclear power plants are using non-renewable resources and creating waste which categorizes this power generation among the unsustainable.

2.2. FUTURE SUPPLY OPTIONS

Often the solutions to the environmental problems from our use of energy are sought only in some environmentally more benign supply options. This is a necessary part of the solution, but only part of it.

Renewable energy sources for producing electricity are in general an order of magnitude less harmful to the environment, than are the non-renewable coal, oil, gas, and nuclear, and hence they will eventually become our energy sources in a sustainable future. Nevertheless, it should be remembered that the renewable resources cannot be exploited at unlimited rates without severe environmental damage. For example, some of the world's first and strongest environmental organizations have been formed to protest against the exploitation of renewable energy sources, namely hydropower. Similarly, some of the worst environmental problems the world is presently facing, are caused by the over-exploitation of renewable energy sources, namely the desertification in many regions due to the use of biomass like trees and bushes as a cooking fuel. Even in Denmark, with widespread use of agricultural and forest waste as a fuel, there are now second thoughts about the future plans for using this resource. The reason is that these plans might be incompatible with a sustainable forestry and agriculture, because a systematic removal of "waste" organic matter can cause a slow deterioration of the soil.

All this points in the direction that in a sustainable energy future we will have to keep energy consumption at a very low level. And electricity is no exception. Fortunately, as we shall see during this workshop, we will not have to freeze in the dark, or otherwise suffer in the future, if we plan in time for a sustainable development. But we certainly have to make significant changes in our development course.

2.3. STORAGE PROBLEMS

Electricity can be stored in batteries, and has been so by electric utilities in the early part of the century as a means to level out peak demand etc. Today, when electricity demand is hundreds of times higher the battery options are not feasible with present battery technologies. It is generally accepted that for the grid electricity has to be produced at the same time as it is being consumed. This is a problem when the demand is fluctuating and even more so when the supply comes from renewable energy sources which are also fluctuating. Expensive reserve capacity has to be available. This is one more reason to hold back on the use of electricity as compared to for instance heat which can be stored more easily.

3. Electricity consumption

Before the oil crisis in 1973, energy demand was included in the energy planning process only as an exogenous input. Demands or consumption were not analyzed, and in general they were not part of the energy planning or the energy policy. Demand was determined more or less by extrapolating passed trends. It was a necessary input to the planning process, but not a variable parameter subjected to a policy. Figure 3 illustrates various prognoses for electricity demand, made around the time of the oil crisis[10]. It demonstrates what happens when demand is not even analyzed in the prognoses, as compared to the actual development in electricity consumption. Gradually this has changed, and during the 1980s concepts like demand side management, DSM, and integrated resource planning, IRP, were introduced[11,12]. They both deal with the inclusion of energy conservation as a policy option.

3.1. ENERGY CHAIN

Electricity is a refined or *secondary* form of *energy*, produced by a supply technology like a power plant or a windmill from some *primary energy* sources like coal, uranium, wind, etc., as shown in Figure 4.

An essential prerequisite for analyzing energy saving potentials is to recognize the fact that *electricity is of no direct use to human beings*. We cannot consume or otherwise enjoy electricity or other forms of energy (excluding the metabolic energy in food). Only by letting the electricity operate some *end-use technologies*, like lamps, refrigerators, washing machines, TVs, pumps, etc., can the electricity be converted into what can be useful, namely *energy services*. These services may consist of illuminating a room, storing food coldly, cleaning clothes, entertaining people, supplying tap water, etc. The conversion of electricity into energy services through end-use technologies is in general very inefficient as demonstrated later in the paper and throughout the workshop.

The principles described here apply not only to the domestic use of electricity, but also to the use in industry, offices and everywhere in society. Electricity end-uses are in general inefficient. While engineers are struggling to gain another two or three *percent* in the efficiency of the power plants, there are no technical or economic problems in achieving a two or three *times*,- or two or three hundred per cent - higher efficiency in the end-use of the electricity.

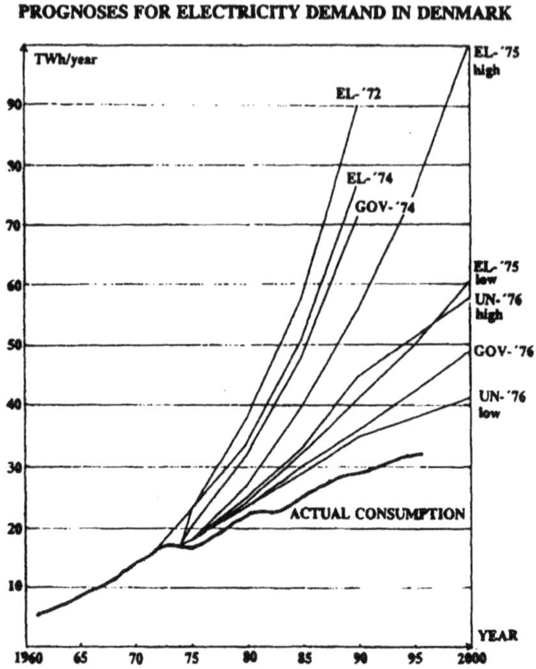

Figure 3.
Various prognoses for electricity demand in Denmark from before demand was analyzed, some made by electric utilities, EL, some by government, GOV, and some by university groups, UN. Lowest is shown the actual development.

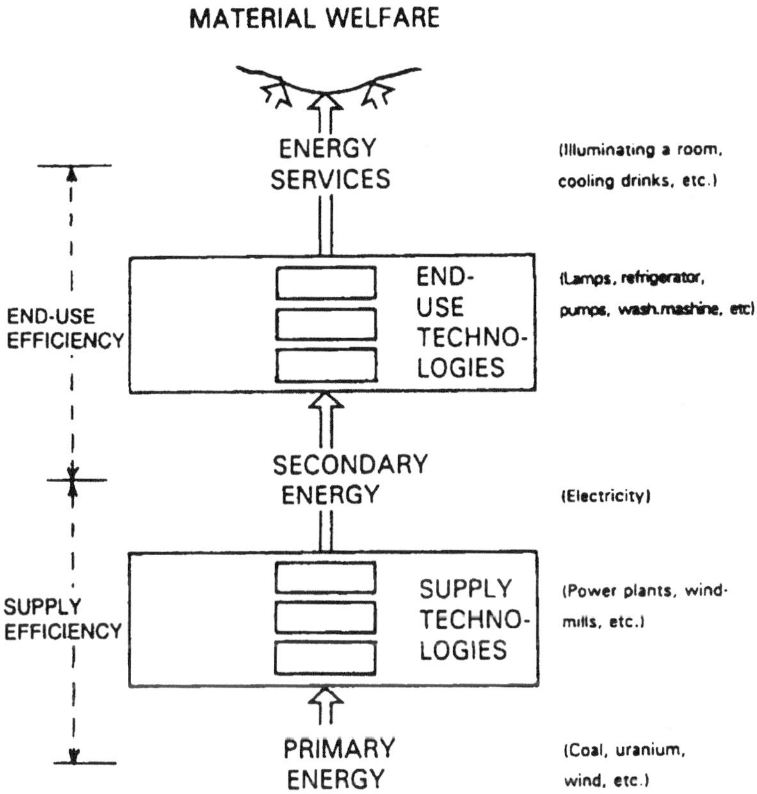

Figure 4.
A chain of technologies converts the primary energy into energy services, here via electricity. As shown here these technologies can be grouped in supply technologies and end-use technologies.

3.2. CONSUMPTION PATTERNS

Since we will have to constrain our use of electricity, it seems like a good starting point to look at the way we use it. For Denmark and Western Europe, Figure 5 shows by the two columns to the right the rather common split-up of electricity consumption according to various economic sectors. The column to the left, however, shows how the same electricity consumption for Denmark is distributed according to various end-uses. These analyses of electricity end-uses have been applied for a study on how much - or how little - electricity is needed in Western Europe if the technological options for using electricity more efficiently are exploited[2].

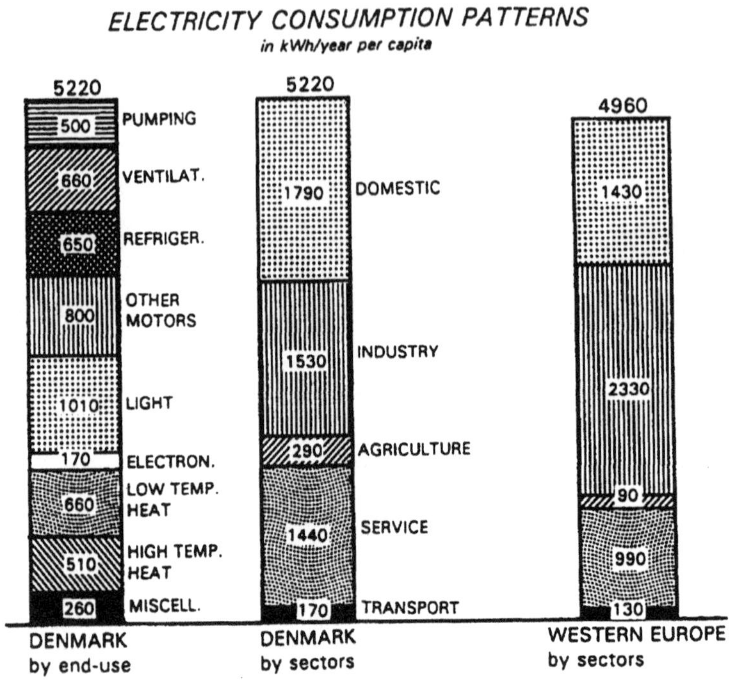

Figure 5.
Electricity consumption patterns in Western Europe and Denmark in 1986, according to end-uses and economic sectors.

3.3. INTERNATIONAL COMPARISON

Figure 6 shows the consumption of electricity per capita in various countries and regions of the world. The differences are striking, and do to a large extent reflect the inequities in general welfare. It is misleading, however, to use the electricity consumption per capita as an indicator of the welfare of people, especially when we look into the future. A society which would ignore the energy saving options discussed here and "waste" its capital on building excessive power plants rather than schools, etc. would rank higher than one opting for efficient use of the electricity.

To the left in Figure 6 is a suggestion shown for electricity consumption which could provide a material living standard close to the present standard in Western Europe if we aim at an efficient use of the electricity[13]. For industry the electricity consumption should here be taken merely as an average, since some countries have electricity-intensive productions of metals, etc. which are then exported as embedded energy to other countries, which can then can have lower consumption statistics on electricity use in industry.

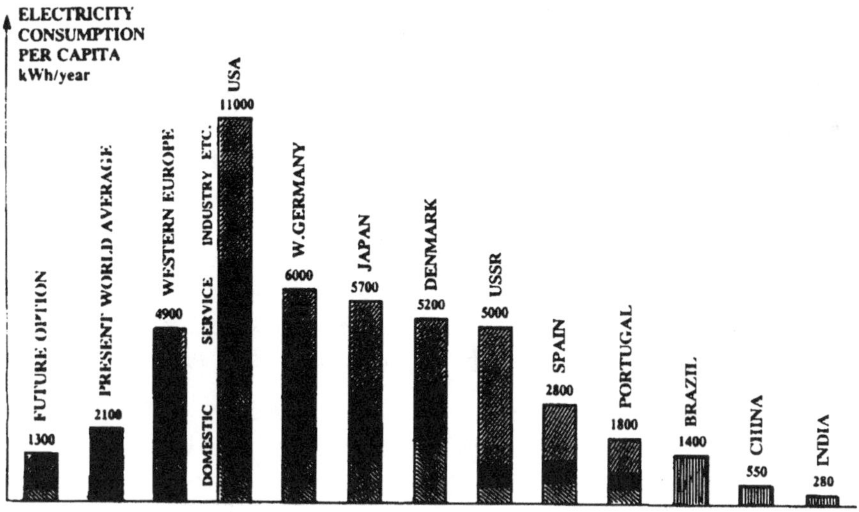

Figure 6.
Electricity consumption per capita in various countries and in Western Europe are here compared to a world average. Also shown to the left is a future option for everybody on earth[13].

4. Technological Saving Options

Above it was suggested that the end-use of electricity is very inefficient. The space in this paper will not allow to go into details, but a few examples will illustrate this, and other sources are referred to in the following.

A general rule for analyzing and for achieving saving potentials is to look at the whole demand system considered. As an example we can take ventilation systems, which typically account for 15% of a country's electricity demand. If we analyze only the electric motor which actually consumes the electricity, we will maybe find a saving potential of 5 to 10%. But starting at the other end, asking what is the energy service we want from the ventilation system, the result will be quite different. Suppose it is a ventilation system for providing fresh air in rooms, experience shows that the air volume necessary can often be reduced, at least in certain hours. A mere 20% reduction in air flow can reduce the need for ventilation power to half of what it was. Better duct systems, fan design, and variable speed control can further cut the need for electricity to run the motor. Saving potentials of 80% are not unrealistic targets in many ventilation systems.

4.1. A LOOK BACKWARDS

Figure 7 shows typical annual consumption of some major electrical appliances in the 1970s, compared with proposals from that time, for how much electricity would be needed to run them in the future, if efficiency was emphasized in the design[14]. Energy planners and decision makers did not take these future options seriously, but *technological* development in the efficiency has been remarkable, nevertheless. It is interesting to notice that models corresponding to the extremely efficient models suggested in the late 1970s are almost all on the market today in one country or another, as indicated in Figure 7 by "Best 1997". This does not mean that electricity consumption has declined correspondingly, because even though the models are on the market, they are not sold very much. The barriers are discussed later in the paper. Today a typical electricity consumption for appliances on the market is roughly around half of what it was in the 1970s.

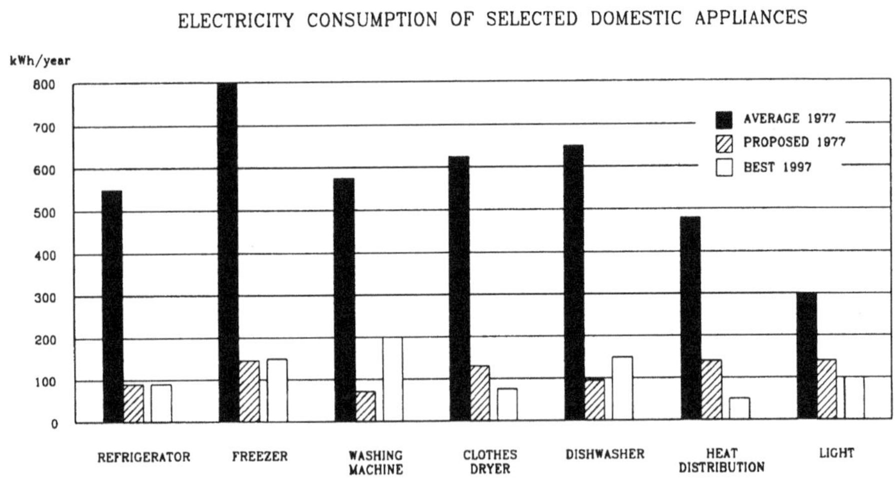

Figure 7.
Typical electricity consumption for major appliances used in 1977 are shown in black. Hatched columns illustrate the consumption suggested at that time as a future option. White columns show the best available models today.

A later study covering two thirds of Denmark's electricity consumption used for lighting and appliances in households and the service sector points towards similar large saving potentials[15], as shown in Figure 8. The figure also shows the various levels of technology and a useful terminology to describe them. It is necessary to be aware that the average used technology, AUS, is often quite different from the average sold technology, AST, since it typically takes 5-20 years to replace the stock of appliances. Similarly, the best available technology, BAT, is not the average sold, since people do not usually choose the best. Future options, which are within reach on the basis of existing knowledge, are here termed efficiency-advanced technology, EAT. The last step shown, EAT+, includes the abolishing of using electricity for inappropriate purposes, for instance providing low temperature heating in washing machines, as discussed in the following section.

More details about the technical potentials for saving electricity can be found elsewhere[16,17,18]. Some of the other contributions to the workshop also provide examples of this. But it is worth pointing out another type of inefficiency in the use of electricity, namely what could be termed the *inappropriate use* of electricity.

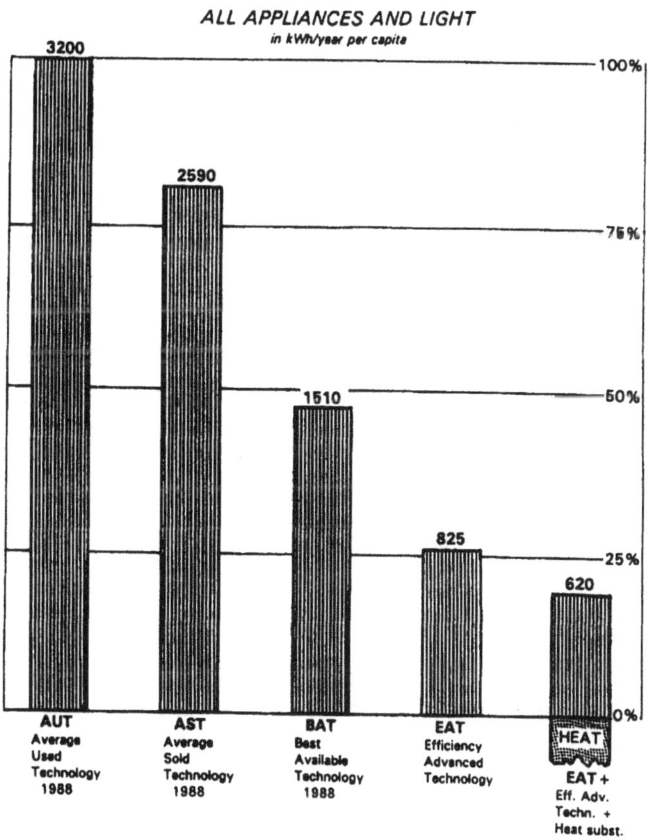

Figure 8.
Potentials for reducing electricity consumption for lighting and appliances, using various levels of end-use efficiencies. The numbers are for Denmark and include households as well as the service sector.

4.2. INAPPROPRIATE USE OF ELECTRICITY

Electricity is a very *high-quality* form of energy, which can be utilized for providing almost any energy service. New uses of electricity will probably be suggested in the future. Some of these uses will make better sense than others. It can make good sense to use more electricity for transport, since here the high thermodynamic quality of electricity, 100% exergy, can be utilized.

When, however, electricity is used directly for low temperature heating purposes, such as space heating or hot water supply (by electric resistance heating), we are dealing with wasteful use of electricity and hence a wasteful use of fuels for generating the electricity. In thermodynamical terms the overall second law efficiency of heating a house by electricity from a normal thermal power plant is only around 2 - 5%. Figure 9 shows the fuel consumption and the CO_2 from providing heat by different systems, when based on the Danish supply system with electricity produced from coal[19]. In Denmark the use of electricity for heating is being phased out through different political actions, as we described in another contribution[20]. Also California has taken steps in that direction. This should be kept in mind in other countries where new power supply capacity is planned and built to satisfy such inappropriate uses as space and water heating.

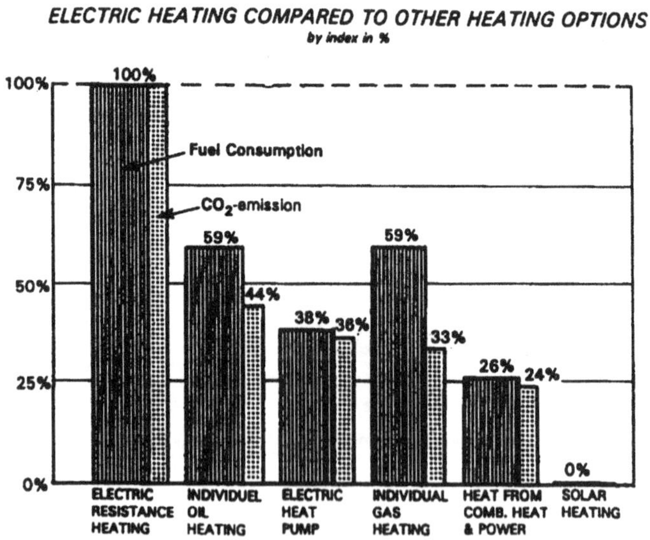

Figure 9.

Fuel consumption and CO_2-emission from heating a house with various fuels and systems.

4.3 ECONOMY OF SAVING ELECTRICITY

Almost anything can be achieved in the area of technology if all cost-effectiveness is ignored. In the real world, however, some considerations have to be made to economic sides.

Considering the various uses of secondary energy for space heating, transport, industrial processes, and for running electric equipment, saving energy seems to be most profitable in the latter category, in the use of electricity. The more energy-efficiently the end-use technologies are designed, the more costly it will be to save another kWh of electricity. It could be argued that the efficiency design changes should be carried to the point where the cost of saving an extra kWh of electricity equals the price of producing a kWh. Since the main incentive for trying to save energy is not to make a profit, but rather to protect the environment, it could also be argued, that the design changes should be carried further, considering the obvious environmental benefits of saving electricity as compared to producing it. For electricity use, the luck is that a long way it is possible to do both, save money and save the environment. Research indicates that more than half of the electricity consumed could be saved cost-effectively.

In calculating the necessary investment for obtaining electricity savings by better technology, two rules are suggested:

Cost of savings should be calculated on the basis of continuing the *natural course of replacement.* Since most electric equipment like household appliances has a life time of 5 - 15 years only, it will usually not make much difference in a long-term energy planning if the replacement is accelerated. When the equipment has to be replaced anyway there will often simply be no extra cost of choosing the most efficient pump system, refrigerator or light system. Or the small extra cost will be paid back in saved electricity within a few months.

Cost of saving should be based *not on the price but on the real extra cost* of making the changed models in the factory. The price is usually determined by the actual demand situation. This means that if a new product is marketed as an energy-saving or environmentally benign model, the manufacturer will often start out by charging a rather high extra price, aiming at the environmentally conscious consumers. Typically the price will be set to give a 2-3 years pay-back time for the buyer. In the long run, which is relevant for planning, the sales price will more likely reflect the extra production cost due to the competition on the market.

For Central and East European countries the opportunities for implementing the efficient end-use technologies in all sectors of societies seem even more profitable than in Western Europe for two reasons: Firstly, the electricity end-use technologies are less energy efficient to begin with. Secondly, a lot of this end-use capital is for other reasons due to be replaced in the near future. It should be recognized, however, that this change in technology is not going to come about automatically by normal, free market mechanisms.

4.4. BARRIERS FOR SAVING ELECTRICITY

A relevant question to ask at this point is why these cost-effective electricity savings are not implemented? What are the barriers for their utilization? There seems to be several barriers, some of which are outlined in the following.

Vested interest and a general ineptness in the power structure of a society plays a considerable role in delaying the electricity saving policy. A future with emphasis on saving electricity calls for different actors than one with continued focus on production of more electricity. This is reflected in the reactions from industry as well as from labour unions, and hence in the policy makers' decisions.

Consumer awareness is very small about the energy saving options. They usually don't care much about energy consumption when a new refrigerator, computer or other appliance is purchased. This is the case for private households as well as in the industry and service sector. One reason for this ignorance is the following.

The *dispersion of electricity consumption* constitutes one of the strongest barriers for saving electricity. It is characteristic that most of our use of electricity is spread out on millions of small uses, each of which seems insignificant, but which together require hundreds of big power plants, emitting millions of tons pollution. Because of this trifle effect alone, huge potentials for saving electricity are often ignored.

The problem of electricity savings, therefore, is not only a technical one. It appears more and more to be a problem of implementation. A *free market* is basically in conflict with saving electricity as well as with other strategies for sustainability, based as they are on collective, long-term considerations. Hence, in order to utilize the benefits of a market economy, it is necessary to accept a *regulated market* as we already have with respect to social issues.

Minimum efficiency standards, energy taxes, subsidies, and other powerful strategies for implementing efficiency should be introduced to regulate the market in a direction of higher efficiency and lower power plant capacity, with all the economic and environmental benefits from that.

In a later paper in this proceedings we will return to barriers and measures to overcome them[9].

5. REFRENCES

1. Goodland, R.: "The case that the World has reached Limits", Chapter 1 in *Environmentally Sustainable Economic Development Building on Brundtland*, ed. by Goodland, Daly, and Serafy. Environment Department, The World Bank, Washington D.C. 1991.
2. Norgard, J.S. and J. Viegand: *Low Electricity Europe - Sustainable Options*, European Environmental Bureau, EEB, Rue de la Victoire 26, bte 12, B-1060 Brussels, Belgium, 1994.
3. Keynes, J.M.: Essays in Persuasion, Macmillan, London, 1931.
4. Mill, J.S.: *Principles of Political Economy*, vol. 2, revised edition, Colonial Press, New York, 1900.
5. Arrhenius, S.: "On the influence of carbon acid in the air upon the temperature of the ground", *Philosophical Magazine and J. of Science*, S 5, 41 No. 251, 237-276, 1896.
6. Daly, H.: "Towards Some Operational Principles of Sustainable Development", *Ecological Economics*, pp. 1-6, 2, 1990.
7. IPCC: *IPCC Second Assessment Synthesis of Scientific-Technical Information*, Intergovernmental Panel of Climate Change, WMO and UNEP, 1995.
8. Danish Energy Agency: *Denmark's Energy Future*, Danish Ministry of Environment and Energy, Copenhagen, 1996.
9. Nørgård, J.S.: "Efficient Technology and Efficient Lifestyles", Chapter in this proceedings, *Development with Sustainable Use of Electricity*, Kluwer Academic Publisher, Dordrecht, The Netherlands, 1998.
10. Danish Government Committee on Analyses and Prognoses for the Energy Sector: *Elektricitetsforbruget frem til år 2000 (Electricity Consumption till the Year 2000)*, Danish Ministry of Commerce, Copenhagen, 1976.
11. Almeida, A.T. and A. Rosenfeld (eds.): *Demand-Side Management and Electricity End-Use Efficiency*, NATO ASI Series E, Vol. 149, Kluwer Academic Publisher, Dordrecht, The Netherlands, 1987.
12. Almeida, A.T., A.H. Rosenfeld, J. Roturier, and J. Norgard (eds.): *Integrated Electricity Resource Planning*, NATO ASI Series E, Vol. 261, Kluwer Academic Publisher, Dordrecht, The Netherlands, 1994.
13. Nørgard, J.S.: "Energy Conservation Through Efficiency and Sufficiency" in *Global Collaboration on a Sustainable Energy Devlopment*, Proceedings from conference in Snekkersten, Denmark, April 1991. Available from Department of Buildings and Energy, Technical University of Denmark, DK 2800, Lyngby, Denmark. 1991.

14. Nørgård, J.S.: "Improved efficiency in domestic appliances", *Energy Policy*, vol.7 no 1, March 1979.
15. Nørgård, J.S.: "Low Electricity Appliances - Options for the Future", p. 125-172 in *Electricity - Efficent End-Use and New Generation Technologies, and Their Planning Implications*, Lund University Press, Lund, Sweden, 1989.
16. Johansson, Th.B., B. Bodlund, and R. Williams (eds.): *Electricity - Efficient End-Use and New Generation Technologies, and their Planning Implications*, Lund University Press, Lund, Sweden, 1989.
17. E-Source, Inc.: *Technology Atlas*, Several Volumes continously updated, E-Source, Inc., 1033 Walnut Street, Boulder, Colorado 80302, USA.
18. Goldemberg, J., Th.B. Johansson, A. Reddy, and R. Williams: *Energy for a Sustainable World*, Wiley Eastern Ltd., New Delhi, India, 1988.
19. Danish Energy Agency: "Varmepumper bør holdes ude fra fjernvarme og naturgassen" (Heat Pumps should be kept out of Regions with District Heat and Natural Gas), *Energinyt* no 3, Danish Energy Agency, Copenhagen, 1993.
20. Gydesen, A.: "Energy Saving Policies in Denmark", Chapter in this proceedings, *Development with Sustainable Use of Electricity*, Kluwer Academic Publisher, Dordrecht, The Netherlands, 1998.

GLOBAL ENERGY EFFICIENCY AND TECHNOLOGIES TO FURTHER ITS PROGRESS

ARTHUR ROSENFELD
EE-1, 6A-081
1000 Independence Ave., SW
Washington D.C.
20585-0121, USA

ALLAN CHEN and ASHOK GADGIL
Lawrence Berkeley National Laboratory
University of California, Berkeley
90-3058,1 Cyclotron Road
Berkeley, CA 94720, USA

Abstract

Since 1973, U.S. gains in the efficient use of energy have saved a remarkable amount of energy and money for consumers and industry, approaching one-half of our energy bills. During a period of scarce energy and high prices that lasted through 1986, the United States doubled the energy efficiency of most new products. Before 1973, U.S. energy was very cheap, and U.S. use of energy per dollar of GNP ("E/GNP") was about twice that of Western Europe (or Japan). The Soviet Union used about four times as much as Europe. Hence U.S. success tells more about the potential in Eastern Europe and the former S.U. than does experience in Western Europe or Japan.

The potential for future savings through energy efficiency is even larger than a second factor of two, through the implementation of proven strategies like revising utility mandates and profit rules to allow utilities to diversify into selling "energy services" and helping customers become more efficient.

Another strategy with a successful record is to encourage the market penetration of more efficient technologies such as:
—Lighting: electronic ballasts, compact fluorescent lamps, mirrored luminaires, occupancy sensors, LED exit lights, etc.
—Windows: Low emissivity (heat mirror) and selective (mirror to both heat and near infrared).
—Motors: variable speed drive, super-efficient motors, substitution of traditional V-belt drive with "cogged" V-belts.

S. Nørgård et al. (eds.), Development with Sustainable Use of Electricity, 19-48.

Promising new strategies are also nearing implementation, including the use of cooler roofing and paving materials and shade trees to reduce air conditioning load and reverse urban heat islands.

The California and New England Utility "Collaboratives."

California has ordered no new central power plants since 1975. Its enviable economic growth has been "fueled" by improved end-use efficiency, small gas turbines, and renewables, mainly wind. To maintain this profitable trend, profit rules have been changed so that utilities become richer by selling energy efficiency than by selling energy.

Reversing summer urban heat islands.

Most cities are summer "heat islands" because vegetation (which cools by evapotranspiration) has usually been replaced with dark-colored roofs and pavements, which get hot under the sun and then heat the air. A study of Los Angeles shows that white roofs, concrete colored pavement and 10 million shade trees will cool the city by 3°C (at 3 pm) and reduce "smog" (ozone) by 12%. Air conditioning savings will be $175,000/HOUR and smog benefits, $360,000/HOUR.

1. Introduction

Since 1973, U.S. gains in the efficient use of energy have saved a remarkable amount of energy and money for consumers and industry, approaching one-half of our energy bills. The potential for future savings is even larger.

Each one percent gain in efficiency means a one percent drop in the need to supply energy (and thus reduces the need to strip-mine or build dams and power plants) and means a one percent drop in the combustion of fossil fuel. The less fossil fuel burned, the longer the delay in the threat of global warming, and the less the pollution by SOx (acid precipitation) and NOx (smog).

Efficient end-use of energy became a matter of public interest in 1973, as a result of the OPEC oil embargo, well before today's strong public interest in environmental protection or climate change. The motivation at that time was to continue economic growth despite shortages of oil and gas and consequent rises in their prices, and there was a public perception that energy prices would continue to rise for the foreseeable future. But the investment in energy efficiency has some enormous, long-term benefits that have outlived the politics of the 1970s: it saves money, stimulates jobs and economic growth, and it also helps the environment.

In the United States, during 13 years of high prices, energy efficiency improved by 35% (measured as GDP/E where E stands for "primary energy"—including the fuel burned at the power plant). Because the economic structure of the U.S. economy during this period

was relatively stable, GDP/E in this case serves as a reasonable indicator of energy efficiency. We choose to discuss the United States, where energy prices were low and energy was used extravagantly, instead of Europe because the largest global environmental challenge today is to contain energy growth in the former Communist block and the underdeveloped nations. These two regions are much more like the inefficient U.S. in 1973 than efficient Europe.
In **Fig. 1**, the solid line is primary energy use and the dashed line is GNP, both indexed to 100 in 1973, the year of the OPEC oil embargo. There are three distinct periods.

Period 1: From 1920 to 1970—low prices. Because energy was cheap, efficiency received very little attention, increasing about 1.0%/year throughout the century (U.S. and most industrial nations). This is not a dramatic enough rise to be visible on the scale of Fig. 1. E and GNP seemed to be inexorably linked. Most economists and politicians, who knew very little science or engineering, believed strongly in what (Rosenfeld) calls the "Wrong Golden Rule Number One," an incorrect rule of thumb of energy economics: "economic growth is linked to energy use." American presidents were more prepared to go to war to protect the nation's "security" interests abroad than to call for more efficient cars and refrigerators, double-glazed windows, and attic insulation.

Period 2: 1970 to 1986—Scarce energy and high prices. Despite considerable confusion in western governments, engineers and scientists began working on efficiency. In 10 years, they doubled the efficiency of almost every new product. Automobile gasoline use dropped from 17 to 8.5 liters per 100 km; refrigerators dropped from 1700 to 850 kWh per year; lighting in new office buildings dropped from 50 to 5 W/m^2, windows went from an average of one or two units of thermal resistance (one for single glazing, two for double) to four, and so on. Beginning with California, the electric resistance heating of water and buildings was forbidden. In only a few major cases did efficiency not double: airplane fuel economy, for example, rose only about 30%. Gas hot water heaters in 1973 were about 60% efficient, but they had pilot lights which reduced their efficiency to 55%, and they have risen since then only to about 85%. The cost of these improvements had a "simple payback time" (SPT) of three years or less, so they were very profitable. SPT is the time required to get an investment back through the savings from reduced energy bills, ignoring the discount rate. The SPT was less than three years in these cases, because in the U.S., consumers and businesses will not buy most services or products unless they pay back in three years or less. The notable exception was oil wells and power plants, which offered an apparently reliable stream of income sufficient to motivate the average investor to tie up capital for 10 to 20 years.

In Fig. 1, which deals with the entire U.S. stock of buildings and equipment, we don't see this doubled efficiency of new products because it has been retarded by the lag between "stocks" and "flows" of goods and buildings. Existing stock has a long service life: cars (10 years), appliances (10-20 years), windows (20 years), buildings (25-50 years). About the time that the efficiency strategy was half completed, it had already saved so much oil and gas that it brought about the collapse of OPEC. President Ronald Reagan and Prime Minister Margaret Thatcher then proclaimed the end of the energy crisis. In one incredible year (1986), President Reagan rolled back auto efficiency standards, vetoed appliance standards, and called for zero funding of efficiency R&D, actions that ushered in the third period.

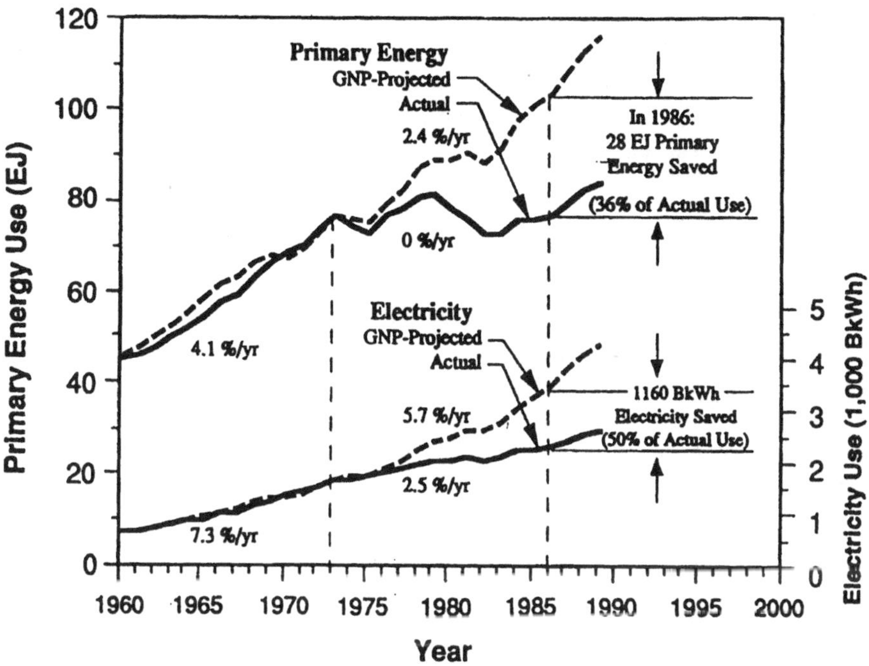

Before the 1973 oil embargo, total primary energy use was growing at the same rate as GNP. Between 1973 and 1986, growth in energy consumption was halted by high oil prices and progressive energy policies.

In 1986, projected primary energy use was 36% higher than actual use, indicating a savings of 28 exajoules. Structural change is credited with 1/3rd of these savings; the remaining 2/3rds, or about 19 exajoules, were saved through increased energy efficiency.

Electricity use followed a similar pattern. Prior to 1973, electricity use was growing roughly 3% faster than GNP. Between 1973 and 1986, growth in electricity use decreased, growing only 2.5% per year, or 3.2% per year less than projected by pre-1973 trends. (GNP-projected energy values are based on 1973 efficiency and GNP. The electricity projections include an additional 3% per year to account for increasing electrification). In 1986, projected electricity use was 50% higher than actual electricity use, indicating a savings of 1160 BkWh.

Electricity use is given in terms of total equivalent primary energy input (exajoules—left-hand scale), and net consumption (1,000 BkWh—right-hand scale).

Energy data: Energy Information Administration, U.S. Department of Energy.
Structural change: Schipper, L., Howarth, R., and Geller, H., *Annual Review of Energy* 15:455-504, 1990 and U.S. Office of Technology Assessment, *Energy Use and the U.S. Economy*, June 1990.

Figure 1.
Total U:S: Primary Energy and Electricity Use: Actual vs. GNP Projected (1960-1989).

Period 3—1986 to present. Cheap prices, minimal R&D, efficiency at near-frozen levels. Here, energy use is rising again, lagging GNP growth by only the traditional 1.0 % per year of 1920 to 1970. In 1988, the threat of global warming became a prominent international issue. The hot, dry summer of 1988 brought the threat of global warming to international attention, but not yet strongly enough.

In our opinion it is prudent to assume that the hot drought of 1988 was partly an early manifestation of global change, but it certainly was not outside the noise. The intervening years were then cooled by the explosion of Mt. Pinatubo. We call your attention to a paper in **Science**, April 7, 1995, by David Thomson. He presents the most convincing evidence to date for global climate change. His evidence includes an advance in the onset of the seasons of up to a week in Europe, and of a few days averaged over the northern hemisphere. The advance tracks well with the increasing concentration of CO_2 in the atmosphere.

The overall message of Fig. 1 is that if society has the will, either through greed (when prices are high) or fear of global warming (when taxes will be high), it can continue to double the energy efficiency of new products every 20 years by several more factors of 2. If it maintains this improvement rate, the old inefficient stock withers, the lag disappears, and world energy efficiency then doubles every 20 years. The 20-year doubling corresponds to a gain of 3.5% per year, which is faster than most economies are growing. A few developing countries are growing faster than 3.5% per year, but they are starting from efficiencies which are twice to four times less than those of the industrial world so they can maintain their growth with only a modest increase in energy supply. That leads to the remarkable and encouraging conclusion that society could **decrease** world energy use by a few percent each year. With a vigorous international family planning program, this would provide the breathing time the world needs to evaluate and respond to the greenhouse threat.

2. Efficiency and the Electric Utilities

Fig. 2 examines the impact of improved efficiency on the U.S. electric industry. It shows that the U.S. economy is growing with a decreasing investment in electric supply. We believe that the healthiest utilities are those which have diversified by investing more in energy efficiency for their customers than in new power plants. There are two "kinks" in Fig. 2 at 1973 and 1983. The one in 1973 represents the cancellation of nuclear plants, and 1983's represents cutting back on new thermal plants. Note that in California, no large thermal power plants have been ordered since 1975; they have simply gone out of style. Fig. 2 has a message for developing countries and those of the former Soviet Union. Of course developing countries need more energy, but Fig 2 shows that energy use can grow considerably more slowly than GDP. This is good news for capital-poor countries.

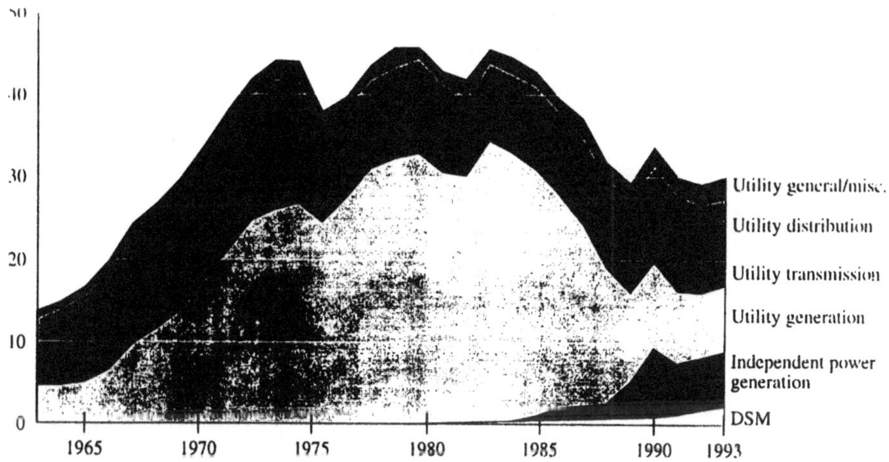

Rise and Decline of Investments in U.S. Electric Capacity. Before 1973, investor-owned utility construction expenditures were growing 14% per year and reached almost $45 billion, close to 15% of total industry investment in plant and equipment. Then nuclear plants began to be cancelled, the Arab embargo spawned an oil/gas shortage, and end-use efficiency rose quickly, reducing the need for further electric growth and leaving the U.S. with overcapacity. Soon even fossil fuel plants were cancelled. Utility investment in generation dropped 7% per year between 1982 and 1992, and no new central plants are planned for the entire West Coast of the U.S.

Source, utility expenditures: Edison Electric Institute, *Statistical Yearbook of the Electric Utility Industry*, 1990 and 1982. U.S. Department of Commerce, *Statistical Abstract of the U.S.*

Source, independent power generation: MW from *Electrical World* (10/89) through 1988; and Edison Electric Institute through 1992; assumed capital cost = $1300/kW.

Source, DSM: Annual EIA Survey Form 861 and *Electric Power Annual*, 1992.

Figure 2.
Investor-Owned Utility and Independent Power Company Capital Expenditures in the U.S. 1963-1993 (billions 1990$).

Revising Utility Mandates and Profit Rules. A century ago, when Thomas Edison started the first U.S. electric company, he sold energy services such as lighting and drive power from motors, not kilowatt-hours. No one knew what a kilowatt-hour was, but customers were eager to buy lighting, motors, and later, refrigeration. Soon society decided that gas and electric utilities were natural monopolies, and should be regulated. Since kilowatt-hours were easier to meter and regulate than the services they provided, the U.S. entered a 100-year regulatory dark age during which its regulatory commissions told utilities to sell only electricity and gas, and set up a profit structure based only on sales of kWh, and gigajoules of gas. Consequently, utilities actually lost money if they encouraged end-use efficiency.

After the success of energy efficiency measures depicted in Fig. 1 and the federal government's abandonment of efficiency in 1986, many regulators, utility officers, and public interest groups decided that the continued improvement of both energy efficiency and economic competitiveness in the United States required a change in the profit rules of utilities to make efficiency more profitable than the sale of energy. One of us (Rosenfeld) was one of a dozen people appointed in 1987 to a task force of NARUC (National Association of Regulatory Utility Commissioners) to invent new rules. The results of this regulatory revolution are shown in Box 1.

Box 1

Statement of Position of the NARUC
Energy Conservation Committee on Least-cost
Planning Profitability

A utility's least-cost plan for customers should be its most profitable plan. However, due to the fact that incremental energy sales increase profits, traditional rate-of-return calculations generally provide substantially lower earnings to utilities for demand-side resources than for supply-side resources. For that reason, profit motive generally encourages utilities to invest in supply-side resources even when demand-side alternatives are clearly identified in its resource plan as being the least-cost resource.

The loss of profits to utilities from relying more upon demand-side resources is a serious impediment to the implementation of least-cost planning. This obstacle to least-cost planning should be addressed. There are identified mechanisms for offsetting the profit-erosion problem.

Therefore, it is the position of the Energy Conservation Committee that state commissions:

1. should require their utilities to engage in least-cost planning;
2. should consider the loss of earnings potential connected with the use of demand-side resources; and
3. should adopt appropriate mechanisms to compensate a utility for earnings lost through the successful implementation of demand-side programs which are a part of a least-cost plan and seek to make the least-cost plan a utility's most profitable resource plan.

(Adopted unanimously by the Committee on Energy Conservation on July 26, 1988.)

The California and New England Collaboratives: In much of the U.S., where electric utilities had ignored the message of Figs. 1 and 2 and had overbuilt, NARUC's suggestion to decouple profits from energy sales was ignored. These utilities *had* to sell kilowatt-hours to pay for their sunk costs in generation and transmission. But the states of the West Coast, and New England had not overbuilt. Changes there came quickly in the form of "collaboratives," agreements among the regulators, the utilities, and other stake-holders. These parties worked together for about a year to design new profit rules permitting the utilities to diversify back into *energy services rather than just energy sales.* One of us (Rosenfeld) personally testified in every state in New England, and participated in the creation of the California Collaborative. Whenever California's largest utility PG&E, does anything to save the customer $1, the customer keeps 85¢, while the remaining 15¢ goes to PG&E's stockholders. That is far more profitable than selling kWh, and so a Kafka-esque change took place. The same utilities, which once promoted electric sales, are now prepared to promote conservation even more creatively. PG&E quickly expanded its conservation program to $200 million (2% of revenues), and started to earn profits of $50-100 million per year, representing a 5 to 10% growth in dividends. Of course the price of electricity had to rise 2% (one time only) to cover the conservation programs and the new stockholder dividends, but these programs are so efficient that PG&E was able to cut to one-third its earlier planned annual $1.0 billion investment in new generation and distribution, an investment that would have raised energy prices inexorably.

Restated in terms of resource planning, the remarkable story is told by **Table 1a**. Before the Collaborative, California utilities had planned to build about 10 GW of new supply in the decade starting in 1991. One GW corresponds to one huge power plant the size of one of the units at Chernobyl, or Three Mile Island, or a major coal-fired plant. Instead, they now plan to avoid building 7 GW of supply, getting the energy services from efficiency, which is three to seven times cheaper. Two-thirds of the remaining 3-GW gap, (2 GW) will be filled by many small "co-generation" power plants, and one-third from renewable sources, mainly wind. A "co-gen" plant is a small gas turbine which "co-generates" both electricity and heat; it is placed near a customer of heat such as a factory, a campus, or a district heating system.

Table 1a. California's Electric Growth, 1991-2001.

	A. GW	B. Savings %/yr.	C. BkWh/yr.	D. $ B/yr. @9.5¢/kWh
Base Case (1990)	46.5	2.0	223	
Total Growth in Demand by 2001 (if California efficiency grows only 2%/yr. like the rest of the U.S.)	10.8	2.0	52	5.0
California's Plan to Meet Demand:				
Improved Electric Efficiency	7.3	2.0 + 1.5	35	3.5
Co-generation and small plants	2.5	0.5	12	1.0
Renewable resources	1.0	2.0	5	0.5
New large central plants	0.0	0.0	0	0.0

California is planning for an increase in the demand for energy services equivalent to 10,800 MW of electric capacity by the year 2001. Of this potential growth in electric demand, about 70% (7,319 MW) will be avoided through demand-side management programs that improve electricity efficiency. This efficiency gain in California is 1.5%/year better than current annual U.S. gain of about 2.0%/year that we assume are included in the base case*. Note (in column D) the remarkable $3.5B/year return on 10 years of utility investment in electric efficiency, which in California is running only a little over $0.3B annually. Of the remaining demand of 3.5 GW of actual new power, about 1 GW will come from renewables. The remaining 2.5 GW will come from co-generation and other small facilities that can use the "waste" heat. *No new central power plants (fossil or nuclear) are planned for California or the entire West Coast during the next ten years.*

Sources:
California Energy Commission, *Electricity Report*, October 1990.
California Energy Commission, *Bi-annual Report*, 1991-1992.
U.S. Department of Energy, *Monthly Energy Review*, April 1992.

California Energy to 2000

* The U.S. base case is derived as follows: Before 1973 electricity consumption divided by gross national product (E/GNP) was increasing about 3% per year. We assume this represented the natural increase in U.S. electric demand. But since 1986, when energy prices dropped, electricity consumption has been increasing only 1% per year faster than GNP, leading us to conclude that efficiency improvements are responsible for the remaining 2% per year drop.

3. Efficiency Elsewhere

In such areas of the world as the former Communist countries and the developing world, efficiency is about four times lower than in the West. There are several reasons. One is subsidized energy prices. Another is lack of access to capital. A third is inefficiencies in economies and planning. (Leaders and bureaucrats in former communist countries and developing nations curry favor with their citizens by subsidizing food and energy. Although it is a market distortion, cheap food prices do not create a serious problem of waste, since people don't overeat just because food is cheap. But low energy prices allow consumers to waste energy, for example, by not sealing their homes to reduce heating loss, a situation that will continue until nations end subsidized energy.)

Figs. 3 [L. Schipper and S. Meyers, 1992] and **4** [J. Martin 1989] compare energy use per capita for the West and for former Communist countries. It shows that, apart from transportation, the USSR in 1987 used five-sixths as much energy per capita as the US, and more than Japan, Denmark, or West Germany, while the USSR produced only one-quarter to one-tenth as much per capita as in the West. Although it is difficult to be precise, one can estimate the energy efficiency of the former Soviet Union to be four to 10 times worse than the West. The optimistic view of this figure is that there is a large potential to improve the energy efficiency of these states.

Not plotted are Poland, Estonia, and Lithuania in 1989, but they were about the same as the USSR. By 1992 the discipline of realistic pricing is reflected in their energy use: the productivity of these states is still low, and homes and buildings are cold, but their energy use/capita has dropped abruptly. Presumably the former Soviet Union will follow.

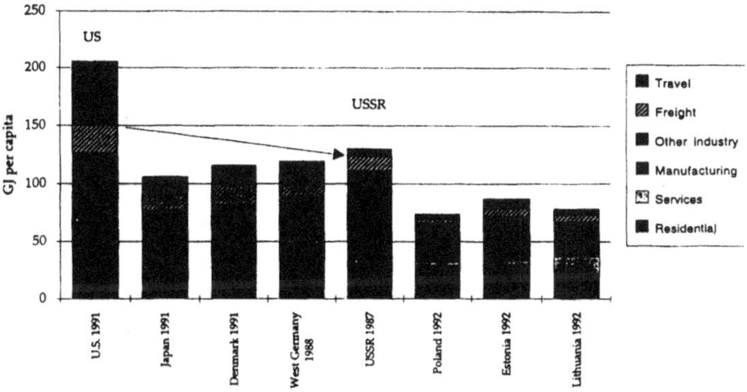

Figure 3.
Final Energy Use per Capita, By sector, 1988-1992.

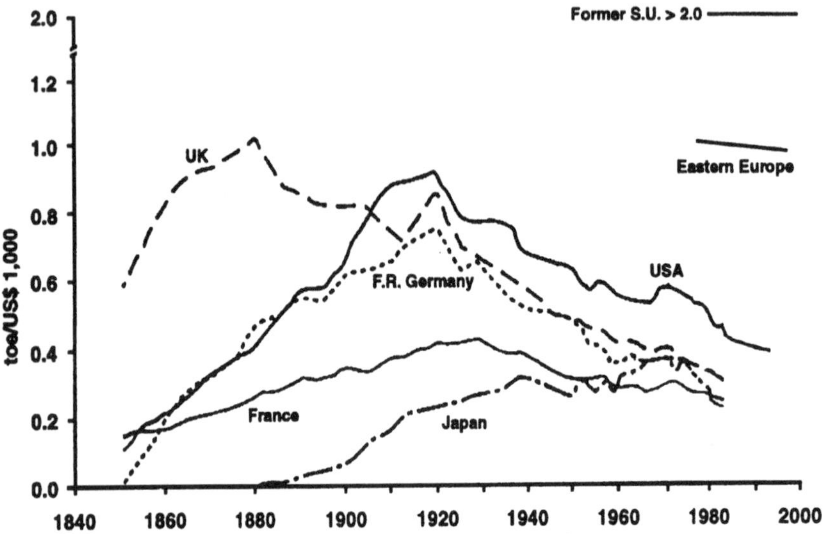

source: Jean-Marie Martin, Revu de L'Energie no. 415, November 1989

Between 1920 and 1970 the U.S. and Germany dropped to 47%, i.e. 1.0%/year. The former Soviet Union estimate of >2 comes from inset figure, when U.S. energy use per capita (without transportation) is comparable with USSR 1987, but GDP per capita differs by at least 4:1.

Figure 4.
Evolution of the Energy Intensity in Different Countries.

4. Efficiency in the former Soviet Union: Replacing Chernobyls with Efficiency

WASHINGTON POST - August 3, 1992

Arthur Rosenfeld and Evan Mills

A Better Idea

An improved light bulb could make dangerous, Chernobyl-class reactors a thing of the past.

Was the Chernobyl nuclear disaster only the first of many? A recent U.N. report raises the ominous prospect of 40 more Chernobyls in the former Soviet Union and Central Europe. A subsequent mishap at the Sosnovy Bor power plant near St. Petersburg reinforced the message. Plainly, the large number of poorly designed, dangerously outdated nuclear power plants in that part of the world poses a major threat to health and safety.

It is a complex problem but one in which a few of the answers are surprisingly simple. Consider just one of them that could have a great impact: a better light bulb.

In the former Soviet Union, lighting alone consumes an amount of energy comparable to that produced by 40 Chernobyls. Much of this lighting is produced in the most energy-wasteful way imaginable. Efficient, compact fluorescent lamps (CFLs) are almost unknown in the former Soviet states. In North America, sales of such lamps will approach 35 million this year and in Europe 90 million. Sales in the West are doubling every couple of years.

CFLs are similar in size and provide the same amount of light as common incandescent lamps, but they consume only one-fourth the energy. At Western electricity prices, they pay for themselves in about six months; Eastern prices have already risen to nearly half of Western. Depending on the extent of their use, these lamps could save the power produced by about five to 10 Chernobyls in the former Soviet Union.

Add to this the opportunities presented by a new generation of efficient long fluorescent tubes that offer about 10 percent savings in non-residential buildings plus substantially better color quality. Running these lamps with efficient and flicker-free electronic ballasts, cleaning dusty and yellowing fixtures and inserting aluminum reflectors to get the same light out of fewer lamps could bring the total savings toward 50 percent. Tally up another five to 10 Chernobyls.

To be sure, there are obstacles: a lack of hard currency, the financial risks that accompany technological change and the challenge of implementation. But there are far more serious obstacles facing those who would try to fix up the existing nuclear plants or find acceptable alternatives to them. In the end, economics tip the scales toward increasing efficiency.

We applaud the West's growing concern about the problem, but the new proposals to spend $10 billion to $20 billion on supply-side fixes are misguided. A less costly approach is to first invest in improved efficiency. For $1 billion, a hundred million CFLs could be diverted to the former Soviet Union from Western markets. Each $1 billion thus spent would turn off five or more nuclear plants.

Alternatively, the West could solve the hard-currency problem by financing construction of, say, 10 CFL factories (at $7 million each) in the former Soviet Union. Electric utilities there or agents from other sectors could moderate the financial risk by guaranteeing to buy the first five years' production (30 million CFLs per factory). Joint ventures between electric utilities in North America and Eastern Europe are already being discussed as a means of marshaling investment in energy efficiency.

In addition to modernizing lighting, about 10 other categories of energy-efficiency strategies are popular in the United States today. Using these, the industry's Electric Power Research Institute estimates a 50 percent across-the-board savings opportunity in this country with a three-year payback time. If our relatively inefficient neighbors to the East were to use the same strategies (something that admittedly would have to be be done with great care) the likely result would be even greater savings.

In any event, a 50 percent savings in the former Soviet Union would equal 150 of their largest power plants. If they wished, the ex-Soviets could then turn off the most dangerous of their nuclear plants, as well as 20 or so of the dirtiest coal-fired power plants. The remaining 100 surplus plants could be used domestically or to sell power to Western Europe for about $25 billion of hard currency each year.

The former Soviet Union is the world's largest producer and consumer of oil (each about 10 million barrels per day), even with its current reduced production. It also has an inefficient economy that uses twice as much energy per unit of GNP as Japan or Western Europe. One result of this inefficiency is that hard-currency oil and gas exports are only about 10 percent of production. With a 50 percent energy-efficiency improvement (in and out of the power sector), exports could zoom to 10 million barrels a day of oil and gas, worth $50 billion each year.

The prospects are bright. Lucrative and innovative solutions seem to be well within reach, if energy planners and policy makers are prepared to recognize the limitations of focusing exclusively on the supply side and realize instead the great gains to be made through greater energy efficiency.

Opportunities for improved energy efficiency are linked to the broader movement toward institutional, managerial and market-oriented reforms and to structural changes that must take place in the former Soviet Union. Making markets for energy-efficient products, fostering competition, pricing energy realistically, metering energy consumption and improving energy statistics are important building blocks.

Averting the impending nuclear nightmare and slowing environmental degradation wouldn't be the only dividends. Opening new avenues for economic development and creating new financial partnerships would increase international security in many ways. We would all sleep easier as a result.

Arthur Rosenfeld is a professor of physics at the University of California at Berkeley and director of the Center for Building Science at the Lawrence Berkeley Laboratory. Evan Mills is the Center's assistant director.

Box 2.
Reprinted in the International Herald Tribune, "Simple Ideas Can Avoid Chernobyls"

4.1. LIGHTING

In the United States, a large and predictable source of energy consumption is the illumination of exit signs. Although this example represents a small fraction of total wattage consumed, it is a striking example of a large opportunity for energy savings through simple technological improvements. Here, the technology for more efficient exit signs already exists in the marketplace. Estimates of the number of exit signs in the U.S. vary widely, but an upper limit is 150 million [Mills, 1993]. A typical exit sign is on during all 8760 hours of the year, and has two 20-watt incandescent lamps, consuming 350 kWh/yr. The lamps generally need to be replaced once a year. Replacing the incandescent with a 3-watt light-emitting diode, energy use falls to 26 kWh/yr. If 150 million incandescent exit signs were replaced with LEDs, the annual energy savings would be close to 50 billion kWh, the output of 10 Chernobyls. Maintenance costs would be lower as well, since the LED-based signs last 10 years or more.

The same argument applies to the former Soviet Union, which has a comparable number of exit signs. These countries are now so poor that most exit signs are not illuminated at all. But as their economies recover, replacing exit signs is a cost-effective way to offset several Chernobyls.

4.2. COGGED AND SYNCHRONOUS V-BELTS

Many motor transmissions, about one-third, are driven by "V-belts" running from pulleys on the crankshaft to the load (e.g. pumps, fans, rollers). However, the conventional V-belt design is not the most energy-efficient one because the belt must flex as it moves around the pulleys. The smooth, flat belt resists bending and generates heat as it bends, thus wasting 5% of full-load power, and about 10% of average power [Nadel, S .et al.1991]. In the 1970s, the industry started making V-belts more flexible by molding them with serrations, producing cogged belts [see **Fig. 5.b.**]. The serrations allow space for the belt to bend, and this cuts the losses by half. Because the belt runs cooler, it lasts about twice as long. Cogged V-belts can replace conventional belts [see **Fig. 5.a.**] on the same pulleys without additional retrofits. Synchronous belts are even more efficient than cogged belts, running at 98 to 99% efficiency. They have no slip because their teeth engage with the teeth of a sprocket pulley [see **Fig. 5.c.**]. The cost of retrofitting a system with new sprocket pulley makes synchronous belts more expensive than cogged belts, although the energy savings are higher.

The cogged design is unknown in the former Soviet Union, even though it costs on the order of only $10,000 to convert a production line from traditional to cogged belts. How much power could this conversion save? Two-thirds of the former Soviet Union's 360 GW drives motors, and one-third of that drives V-belts—about 80 GW. The reduced losses from cogged belts will save up to 5% of this, about 4 GW (4 Chernobyls)—

providing a saving whose first cost is a few tens of thousands of dollars—and reduced life-cycle cost as well. Unfortunately, international organizations like the World Bank, the European Bank for Reconstruction and Development and many others have taken too long to fund, or just guarantee, investments in such simple ideas as efficient exit signs and cogged V-belts.

Pirelli — Oleostatic® Conventional V-Belt Drives
A,B,C,D,E

Pirelli — Cogged Conventional V-Belt Drives
AX,BX AND CX

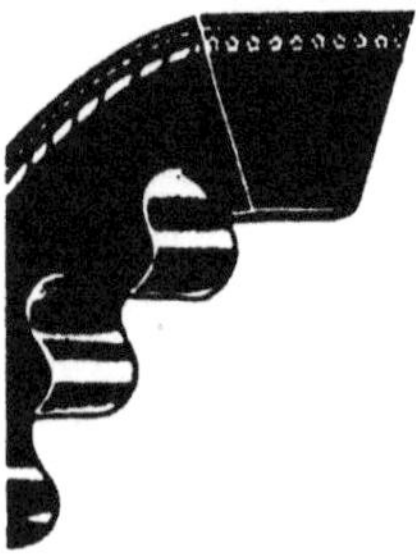

Pirelli Linea-X Cogged Raw Edge Conventional Belts have molded cogs for increased flexibility and reduced stress due to bending. Cogs are designed to dissipate heat for cooler running drives with less internal heat build-up.

Cogged Conventional — A More Efficient Drive Package
- Greater Flexibility — Saves Energy
- High Horsepower Capacity — Fewer Belts and Sheave Grooves
- Increase Drive Life — Less Downtime
- Better Grip — Less Slip, Energy Saving
- Reduced Weight — Reduces Drive Cost, Saves Energy
- Higher Drive Rates — Smaller Sheaves

Cross Sections

Pirelli Designed and Produced for Performance

① Synthetic Fabric Top — Oil resistant with a special woven process to increase flexibility.

② Synthetic Tension Member — Load carrying is designed to remove stretch, improve shock resistance and increase flex life.

③ Fiber Reinforced Cushion — Heat resistant, compounded for full support and superior flex life.

④ Fully Molded Cogs — Provide flexibility and cord support.

⑤ Raw Edge — Long wearing, oil and heat resistant, high co-efficiency of friction reduces slip and ensures aggressive performance.

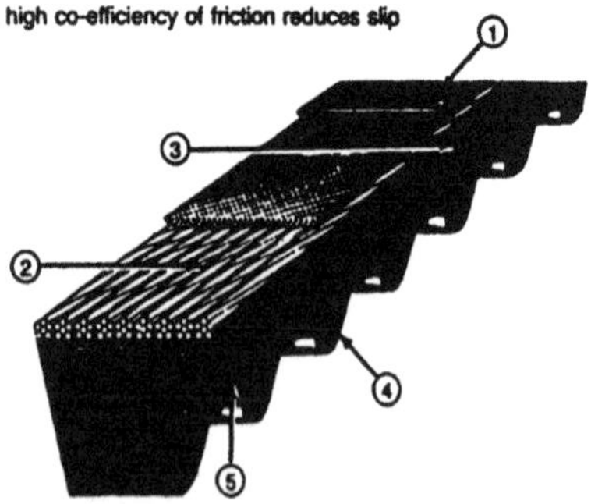

Figure 5.a. 5.b. 5.c.
Examples of Conventional and Cogged V-belts.

5. Summary of savings from technologies developed at Lawrence Berkeley Laboratory

Since the oil price shocks of 1973, Lawrence Berkeley Laboratory has pioneered the development of technologies and policies to save energy and benefit both energy users and the economy. Among the innovations whose development and commercialization were assisted by LBL are the electronic ballast for fluorescent lamps (replacing magnetic ballasts) and energy-efficient window coatings for double and triple glazings. In **Table 1b**, energy savings from three technologies developed in association with LBL, electronic ballasts, compact fluorescent lamps and low-emissivity coatings, are described in columns A, B and C, and the effects of one type of energy efficiency standard, for refrigerators and freezers, are tallied in column E. LBL is the primary contractor charged with developing recommendations for appliance standards for the Department of Energy.

Col. A Electronic "ballasts" have a 25% marketshare of all U.S. ballast sales, expected to rise to 75% by 2015. All fluorescent lights require 'ballasts" to control the power flowing through them. The older, 60-Hz magnetic ballasts flickered, could hum and were inefficient. Thanks to LBL's efforts beginning in the late 1970s with ballast manufacturers and local utilities, the electronic ballast is a growing building technology that has saved $750 million in energy costs by 1993, growing to $6 billion/year net by 2015. At this point, electronic ballasts will save energy equivalent to about 16 huge (one-GW) baseload power plants (assuming that one plant supplies about five billion kilowatt hours per year) or the equivalent of 45 offshore oil platforms.
Col. B Although the electronic ballast is major energy saver, the compact fluorescent lamp (CFL) has a larger potential to reduce energy use—CFLs are four times more efficient than incandescents. One CFL saves 130 liter of gasoline equivalent, enough to drive 1750 km. In the U.S., when CFLs reach market saturation, they are expected to save about 110 billion kilowatt hours annually, or the equivalent of about 22 baseload power plants or some 60 offshore oil platforms.

Col. C A square meter of energy efficient window saves the equivalent of about 300 liters of gasoline, or, at 7.5 l/100 km—3500 km, about enough to travel from Porto to Hamburg. Energy-efficient "low-emissivity" window coatings now have a 36% marketshare of all windows, and they are expected to grow to 80% by 2015. The Laboratory's effort began in 1976. An expenditure of $3 million (through the early 80s) of research dollars has increased window industry revenues of $630 million per year (the incremental cost of so-called low-emissivity windows over the cost of an equivalent number of standard windows). This also compares with the cumulative energy savings through 1993 of $3 billion, growing to $4 billion annual savings (net of up-front costs paid by consumers) by 2015. When the market is saturated, energy-efficient window coatings are expected to save the energy equivalent of about 35 large offshore oil platforms, assuming each platform produces 10,000 barrels of oil per day.

Col. E Appliance standards have moved cost-effective advanced technologies in the marketplace. LBL develops appliance standards for the U.S. DOE, mandated by the National Appliance Energy Conservation Act, which was passed by the U.S. Congress in 1986. The 1992 California standards for refrigerators and freezers at saturation will provide an annual energy savings of 130 billion kWh, equivalent to $9 billion. This represents a savings of 26 one-GW power plants or 70 offshore oil platforms. The simple payback time for the California refrigerator standards is five years. Standards have been proposed or are under development for other appliances, including dishwashers, clothes washers and dryers, lamps, ballasts and water heaters. They will accelerate the market penetration of advanced new energy-efficient appliances based on emerging technologies that, in some cases, already exist but are just starting to be commercialized, and that will have other benefits for consumers unrelated to energy use. All NAECA standards together could reduce U.S. energy use 23 Quads by the year 2015, equivalent to 1.4 years of U.S. residential energy use, and a reduction of 400 million tons of carbon emissions.

Col. D Tools for building designers include an advanced computer program called DOE-2 that was begun in 1977. Users of DOE-2, which is distributed as public-domain software, identify savings worth $85 million per year in energy bills, adding up to $1.9 billion for structures designed through 1993 with the program. DOE-2 analysis services represent a $35 million annual business for consultants. The growth in PC processing power, the ubiquity of PCs in the design industry, and the designers' and utilities interest in saving money for their clients offer an opportunity to spread software that embodies energy-efficient building design principles.

The energy estimates described in Table 1b may be underestimated. Column C shows a premium of $10/m^2$ for low-emissivity windows, but this does not account for the savings from downsized air conditioners for windows installed in warm climates. Air conditioners use less energy thanks to reduced solar heat gain through these windows. In these cases, the savings in air conditioning exceeds the cost the premium paid for low-E widows, and the first cost of the building drops. This effect is most impressive in automobiles, where an assumed premium of $50 to add low-E film to an already laminated windshield can halve the size of the air conditioner. Thus, over the life of the car, the owner saves energy from two sources: the car's reduced weight, and from the reduced power needed to drive the air conditioner—up to several percent in fuel [Hopkins 1994].

Table 1.b. Economics of Three New Energy Efficiency Technologies and Appliance Standards. A 1994 tables 1 & 4 of Geller et al., *Annual Review of Energy* 1987, Vol 12.

	RESEARCH & DEVELOPMENT				STANDARDS
	A HIGH-FREQUENCY BALLASTS VS. CORE COIL BALLASTS	B COMPACT FLUORESCENT LAMPS [1] VS. INCANDESCENTS	C LOW-E (R-4) WINDOWS VS. DOUBLE GLAZED WINDOWS Per small window (10 ft^2)	D TOTAL	E REFRIGER-ATORS AND FREEZERS '76 base case vs. '92 CA Stds.
1. UNIT COST PREMIUM [2]					
a. Wholesale	$8	$5	$10		$100
b. Retail	($12)	($10)	($20)		($170)
2. CHARACTERISTICS					
a. % Energy Saved	33%	75%	50%		66%
b. Useful Life [3]	10 years	3 years	20 years		20 years
c. Simple Payback Time (SPT) [4]	0.8 year	0.5 year	2.9 years		1.3 year
3. UNIT LIFETIME SAVINGS					
a. Gross Energy	1330 kWh	440 kWh	10 MBtu		20,720 kWh
b. Gross $ [5]	$100	$33	$70		$1550
c. Net $ [3b-1a]	$92	$28	$60 [6]		$1450
d. Equivalent kg coal [7]	500	175	300		8,000
e. km in car @ 7.5 liters/100 km	5000	1750	3500		80,000
4. SAVINGS 1985-1993					
a. 1993 Sales	25M	42M	20M		6M
b. Sales 1985 through 1993	54M	147M	90M		50M
c. Cum. Net Savings [4b x 3c]	$750 M	$1.4B	$3B	$15B/8yr	$73B
5. SAVINGS AT SATURATION [8]					
a. U.S. Units..	600M	750M	1400M		125M
b. U.S. Annual Sales	60M	250M	70M		6M
c. Annual Energy Savings [5b x 3a]	80 BkWh	110 BkWh	0.3 Mbod		130 BkWh
d. Annual Net $ Savings [5b x 3c] [9]	$6B	$7B	$4B	$17B/yr	$9 B
e. Equivalent power plants [10]	16 "plants"	22 "plants"		38	26 "plants"
f. Equivalent offshore platforms [10]	45 "platforms"	60 "platforms"	35 "platforms"	140	70 "platforms"
g. Autos offset [11]	16 M	22 M	12 M	50 M	26 M
6. PROJECT BENEFITS					
a. Advance in Commercialization	5 years	5 years	5 years		5 years
b. Net Project Savings [6a x 5d]	$28 B	$35B	$21B	$84B	$45 B
7. COST TO DOE FOR R&D	$3M	$0 [12]	$3M	$6M	$2M
8. BENEFITS/ R&D COST [6b/7]	9,000:1		7,000:1	14,000:1	23,000:1

From: "The Role of Federal Research and Development in Advancing Energy Efficiency," Statement of Arthur H. Rosenfeld before James H. Scheuer, Chairman, Subcommittee on Environment, Committee on Science, Space, and Technology, U.S. House of Representatives, April 1991. Available from Center for Building Science, LBL, (510) 486-4834.

---

(1) Calculations for CFLs based on one 16-watt CFL replacing thirteen 60-watt incandescents, burning about 3300 hours/year, assuming that a CFL costs $9 wholesale, or $5 more than the wholesale cost of thirteen incandescents. For retail we take a lamp cost of $18.

(2) Unit cost premium is the difference between one unit of the more efficient product (e.g. one high-frequency ballast) and one unit of the existing product (e.g. one core-coil ballast).

(3) Useful life is the assumed calendar life of the product (as opposed to operating life such as burning hours for a lamp) under normal operating conditions. A commercial use is assumed for CFLs, but labor savings are not included.

(4) SPT is the number of years required to recoup the initial incremental investment in an energy-efficient measure through the resulting reduction in energy bills.

(5) Assuming price of 7.5¢/kWh for commercial sector electricity and a retail natural gas price of $7/MBtu (70¢/therm).

(6) For hot weather applications where low-e windows substantially reduce cooling loads, air conditioners in new buildings can be down-sized, saving more than the initial cost of the low-e window.

(7) 1 kWh - 0.4 kg coal For windows, use 10 MBtu = 10 GJ= 30 l of petroleum or gasoline

(8) Saturation is 100% of the market for all products excepts CFLs. It is unrealistic to assume that CFLs will replace infrequently used incandescents; thus, we have defined market saturation for CFLs as 50% of current energy used by incandescents.

(9) Net annual savings are in 1990 dollars, uncorrected for growth in building stock, changes in real energy costs, or discounted future values. See Geller et al., Annual Review of Energy, 1987, Vol. 12, Table 1. Note that we attribute energy saved by the product over its useful life to the year it gets sold.

(10) One 1000 MW baseload power plant supplying about 5 BkWh/year = 57 x 10^{12} Btu = 0.1x Alaskan Arctic National Wildlife Refuge (ANWR). One offshore oil platform = 10,000 bod. To convert "plants" burning natural gas to "platforms": 1 "plant" = 27,000 bod = 2.7 "platforms." ANWR, at 0.3 Mbod, is equivalent to about 30 "platforms."

(11) 1 automobile (400 gallons/year) generates 1 tonne carbon per year. Thus electricity and gas savings can be converted to "autos offset" (1000 MW power plant is equivalent to 1 M autos).

(12) Descended from high-frequency ballasts (only DOE assistance was in testing).

6. Cooling Communities with Cool Roofs, Cool Roads, and Shade Trees

Modern urban areas usually have darker surfaces and less vegetation than their surroundings. These differences affect the climate, energy use, and habitability of cities. At the building scale, dark roofs are heated by the summer sun and thus raise the summertime cooling demands of buildings. Collectively, the hot dark surfaces transfer heat to city air, and vegetation, which cools the air through evapotranspiration, is relatively absent. These effects warm the summer air over urban areas, leading to the creation of the summer urban "heat island." On a clear summer afternoon, the air temperature in a typical city is about 2.5°C (5°F) hotter than the surrounding rural area (**Fig. 6**). Akbari et al. [1992] have found that peak urban electric demand in six American cities (Los Angeles, CA; Washington DC; Phoenix, AZ; Tucson, AZ; and Colorado Springs, CO) rises by 2 - 4% for each 1°C rise in daily maximum temperature above a threshold of 15° to 20°F (Los Angeles case shown in **Fig. 7**). Thus, the additional air-conditioning use caused by this urban air temperature increase is responsible for 5-10% of urban peak electric demand, at a direct cost of several billion dollars annually.

The Heat Island/Cool Communities Project at the Lawrence Berkeley Laboratory has examined both the building- and city-scale effects of the urban surface on energy use and climate. At the building scale, cool roofs reduce air conditioning load. At the community scale, they also reduce smog. Numerous experiments on individual buildings in California and Florida show that painting the roof white reduces air conditioning load between 10 and 50%, depending on the thickness of insulation under the roof. At the community scale, heat islands research at LBL finds that increasing the albedo[1] (solar reflectivity) of urban surfaces and planting trees in urban areas can limit or reverse the urban heat island effectively and inexpensively. Both of these improvements can be implemented by 1) rating and labeling roofing materials by their temperature rises on a cloudless summer day; 2) adopting relatively mild standards (for example, that new roofs run cooler than halfway between the surface temperatures of typical white and black surfaces) and 3) offering rebates on new roofs (or re-roofs) for beating the standards. Who would be responsible for these three measures?

---

[1] Albedo is defined as hemispherical reflectivity integrated over the solar spectrum. Low-albedo surfaces absorb a larger portion of the incident insolation and become hotter than high-albedo surfaces. Most high-albedo surfaces are light-colored, although selective surfaces which reflect a large portion of the infrared solar radiation but absorb some visible light may be colored, yet have relatively high albedos.

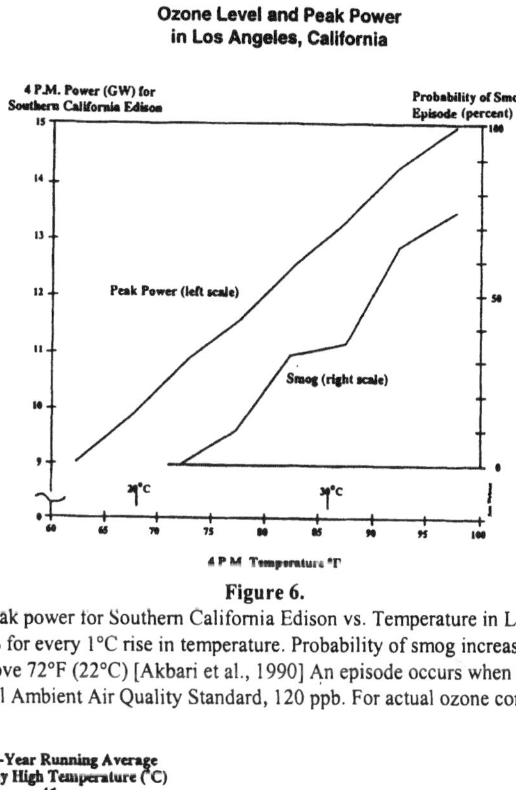

Figure 6.
Ozone levels and peak power for Southern California Edison vs. Temperature in Los Angeles, CA. Peak power use rises by 3 % for every 1°C rise in temperature. Probability of smog increases by 6 % for every 1°C rise in temperature above 72°F (22°C) [Akbari et al., 1990] An episode occurs when the ozone concentration is above the National Ambient Air Quality Standard, 120 ppb. For actual ozone concentration, see fig. 6.

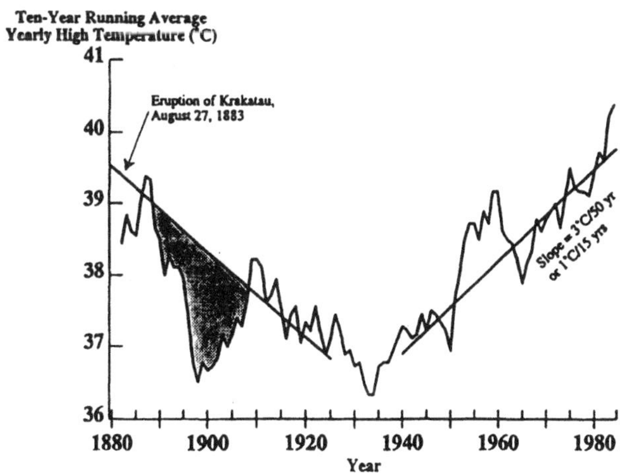

Figure 7.
Ten-year running average high temper<tures in downtown Los Angeles, CA (1882-1984). With increasing irrigation and orchards, Los Angeles cooled 2°C/year until the 1930s. Then, as asphalt replaced trees, Los Angeles warmed 3°C (6°F). The pronounced temperature depression in the late 1880s-90s is due to the eruption of the Krakatau volcano.

1. Temperature ratings. In the US, the roofing and roadway industries are working with ASTM (American Society for Testing Materials) to define a "Relative Coolness Index" rating, and to adopt a Relative Coolness label for their products.
2. The California Energy Commission will issue an extension of its new building performance standard taking roof temperature into account. ASHRAE (the American Society of Heating, Refrigeration, and Air Conditioning Engineers) has a "Building Standard Series 90" committee which is studying adding roof temperature.
3. In California stockholders of a utility share in about 10% of savings which the utility achieves for its customers. If a home-owner who needs a new roof anyway switches the color of the replacement roof from brown to white, her peak power demand will drop by about 1 kW, and her annual air conditioning bill will drop by $50-100. Thus the utility can avoid an investment of about $1000 in new energy supply, and can easily afford a $500 rebate. It's stockholder profit about $50. New, whiter roofing materials are appearing in the market, and their prices are negligibly higher.

Fig. 8 shows the midday temperatures of various horizontal surfaces exposed to sunlight. For highly absorptive (low-albedo) surfaces, the difference between the surface and ambient air temperature, may be as high as 50°C (100°F), while for less absorptive (high-albedo) surfaces, such as white paint, the difference is about 10°C. For this reason, shade trees (i.e trees which directly shade buildings) and cool surfaces (which absorb little of the incident insolation) are effective means of direct cooling and reducing energy use. Through direct shading and evapotranspiration, trees reduce summer cooling energy use in buildings at about 1% of the capital cost of avoided power plants plus air-conditioning equipment [Akbari et al., 1990]. Cool surfaces are more effective than trees, and cost little if color changes are incorporated into routine maintenance schedules. Also, the results from light-colored surfaces are immediate, while it may be ten or more years before a tree is large enough to produce significant energy savings. Akbari et al. [1989] discuss the relative benefit/cost of white surfaces vs. trees.

LBL researchers have simulated the cooling achieved by increasing the reflectivity (albedo) of roofs and roadways in the Los Angeles Basin (Sailor and Kessler 1993). About 40% of the area of the urbanized "cells" in the basin are covered by roofs and roads which can realistically have their albedo raised by 30% at the time that roofs and roads receive their normal maintenance. The results are shown in **Fig. 9**: a 2 °C (4°F) cooling by noon, when the smog is forming rapidly. Putting these results into the Los Angeles smog model then predicts a reduction in population-weighted smog of 10-20%. To achieve this gain by removing cars, or electrifying them, would require removing about 3 million cars. Adding shade trees to these simulations increases these gains by another 50%. And these results will save money because a cooling of 3°C at 3-4pm will reduce peak power consumption in Los Angeles by 0.6 to 1.2 GW (worth at least $100,000/hour).

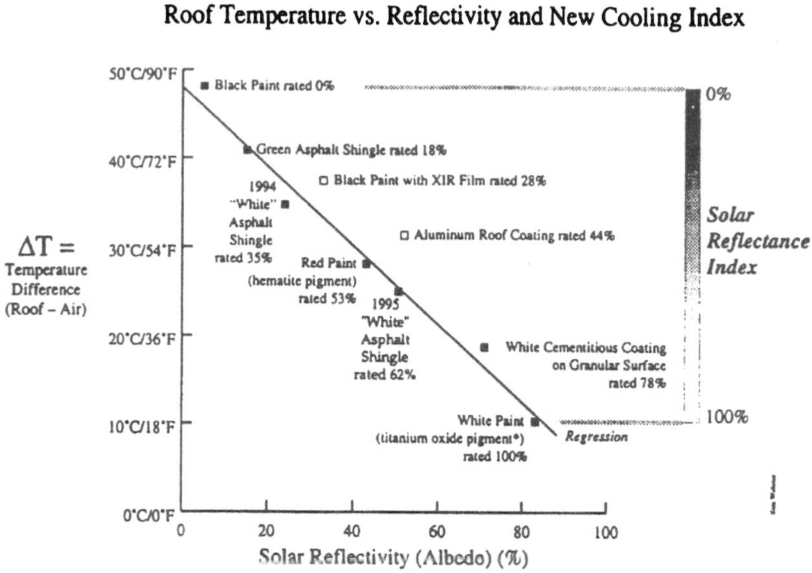

Figure 8.
The difference between surface and air temperatures vs. Solar absorptivity of paints and roofing materials facing the sun. Note that white is 40°C cooler than black. Metals (open squares) become hotter than similar colored paint because they radiate poorly (low emissivity). Before 1960, white shingles were cooler, but then industry made them darker to hide dirt. White acrylic paint is pigmented with titanium dioxide. Open squares are not included in the regression. Solar reflectivity is measured according to ASTM E903. Data from Berdahl and Bretz (1994).

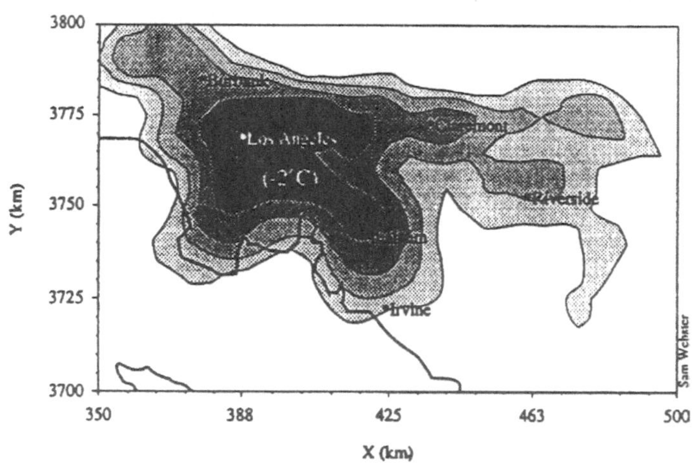

Figure 9.
Simulation of temperature rise at noon when average albedo is raised by 13 % in the urbanized cells of Los Angeles basin. At noon, when the smog is "cooking," a large area has cooled 2°C. By 3 pm the cooling reaches 3°C

TABLE 2. Base case U. S. air-conditioning use and savings potential of cool surfaces and shade tree program assuming 20% of air conditioning is avoided by 2015. (some figures are rounded).

	Year	1995	2000	2005	2010	2015
1	Base Case U.S. A.C. Use					
1a.	Electricity (BkWh)	441	464	488	512	539
1b.	\$(utility + customer) cost†	\$44 B	\$46 B	\$49 B	\$51 B	\$54 B
1c.	CO_2(MtC*)	110	116	122	128	135
2	Annual Savings					
2a.	Fraction of Base Case (%)	0	5	10	15	20‡
2b.	Electricity (BkWh)	0	23	49	77	108
2c.	\$(utility + customer) cost†	0	\$2.3 B	\$4.9 B	\$7.7 B	\$11 B
2d.	CO_2(MtC)	0	6	12	19	27

† Assuming 1 kWh costs 10 cents in 1994 dollars.

* MtC= million metric tons of carbon

‡ Potential savings in 20 years when roof resurfacing is completed and trees have matured.

7. Policy Steps to Implement Cool Surfaces and Shade Trees Program

Table 2 describes the potential savings. However, achieving this potential is conditional on receiving the necessary Federal support. Programs for planting shade trees already exist, but to start an effective and comprehensive program requires the following eight steps. American Forests, our outreach collaborators, will share responsibility for steps 5,6, and 7.

1. Work with ASTM and industry to create test procedures, **ratings**, and labels for cool materials.
2. Assemble a cool materials **data base** made widely available to industry, utilities, contractors, architects, roofers, state and local procurement officers, consumers and communities.
3. Incorporate cool roofs and shade trees into the **Building Energy Performance Standards** of ASHRAE, CABO, California Title 24, and Air Quality Management Districts. Standards can be relatively mild if accompanied by Step 4.
4. Offer **utility rebates** or other incentives to beat the standards. This will require support by the state public utility commissions.
5. Begin **information programs** for all the groups mentioned in Step 2, and distribute information by grassroots support networks to building owners and local governments.
6. Demonstrate savings in selected **"Cool Communities,"** including Federal facilities, particularly military bases. This will require support by the local utility.
7. Establish aggressive policies for the **procurement of cool roofing materials** by Federal, state and local governments. Create "purchasing co-ops" in the Cool Communities.

8. Expand the Los Angeles Basin "RECLAIM" NOx-credit trading market to include air temperature reduction by cool surfaces.
   The California South Coast Air Quality Management District plans to monetize any growth of smog above a "ceiling," so that smog feedstocks (specifically NOx) can be traded on a "credit stock-market" to be called "RECLAIM." An expanding industry or utility that really needs to produce more NOx can purchase NOx credits on the RECLAIM market from another producer who has a relatively lower cost for reducing NOx emissions.
   SCAQMD and EPA now recognize that air temperature is as much a cause of smog as NOx or volatile organic compounds, so that cool surfaces and shade trees should be monetized on RECLAIM along with NOx. In this case a cooler roof might be worth $1000. The Cool Communities program will work with EPA, SCAQMD, and its counterparts in other smoggy cities, to model the dependence of smog on cool surfaces and vegetation, so as to calculate the trade-off ratios between, for example NOx and a cool parking lot.

8. Stopping cholera with UV lamps, while saving firewood.

In the developing world, waterborne diseases such as cholera, typhoid fever, gastroenteritis, dysentery, and infectious hepatitis kill more than 400 children every hour, and result in the loss of billions of hours of worker productivity each year. Municipal tap water is uncommon in many developing country households, and two out of three people in the world must fetch water from outside their homes. In India, water purity issues are particularly important during the monsoon season when heavy rainfall washes raw sewage and other contaminated material from the fields into the wells and surface water. Disinfecting water by boiling it with firewood puts extreme stress on the supply of firewood in many countries and reduces fuel available for cooking. Gathering the wood occupies time that might be spent productively in other activities.

In order to address this significant public health problem, there is an effort underway at the Lawrence Berkeley Laboratory (LBL) to introduce a water purification system using ultraviolet light to rural villages in India, Mexico and South Africa. The goal of this project is to design and field-test a water purification device for developing countries that is durable, easy to use, inexpensive, and can be constructed and maintained locally.

The team began its research early in the summer of 1993 and increased its efforts considerably in August 1993, when an outbreak of cholera was reported in India, Thailand, and Bangladesh [Altman, 1993]. A year later, the cholera epidemic continued to be a problem in India—in the state of Bihar, between the months of May and August 1994, approximately 2,200 people died from cholera [Times of India, 1994]. Other waterborne diseases also pose a serious health threat to Indian communities. In the state of Orissa alone, there are approximately 300 infant deaths per day as a result of waterborne gastrointestinal diseases [Alkari, 1994].

The researchers estimate that the UV disinfection system can provide clean drinking water for less than 5¢ per villager annually. The disinfection process is highly energy-efficient and **uses approximately 40,000 times less primary energy than the standard alternative — boiling water over a cookstove**. The provision of a simple and inexpensive method for disinfecting drinking water will save the lives of many people, particularly the lives of children, who are the most susceptible to diarrheal diseases. Because women are primarily responsible for providing their families with water as well as bearing and caring for children, the UV disinfection system has the potential to greatly improve women's quality of life by reducing their workloads as well as the number of children they lose to waterborne diseases.

9. The Ultraviolet Water Purification System

The technology uses ultraviolet (UV) light to kill waterborne pathogens (bacteria, viruses, and molds) from the local water supply. Ultraviolet light is classified by three different wavelength ranges.[5] UV-C light is "germicidal"; that is, it destroys bacteria, viruses, and other pathogens by inactivating their DNA and thus their ability to reproduce. Light with a wavelength of 254 nm gives the highest germicidal efficacy in the ultraviolet range. Because this is the wavelength at which a low-pressure mercury vapor lamp emits roughly 90 percent of its light, the standard fluorescent lamp technology can be used in the disinfection system.

The glass tubes of the fluorescent lamps that light our offices, and sometimes homes, are coated with a phosphor that absorbs ultraviolet light and gives off visible light. The lamp that is used in the UV disinfection system is similar to a standard fluorescent lamp, but the lamp tube is not coated with phosphor and is made of a special glass that is transparent to UV light. This "germicidal" variety of lamp is already manufactured by many large companies that make standard fluorescent lamps. Consequently, lamps, ballasts, and starters for the UV disinfection system can be bought off-the-shelf, with the full benefits of mature volume production (i.e. at low cost and free of technical bugs).

The estimated one-time capital cost of an ultraviolet system is $300, including valve, fittings, and labor. The life of the PVC chamber is expected to be approximately 40 years; the UV lamp requires replacement annually. Assuming that the system is

---

[5]UV-C from 100 nanometers (nm) to 280 nm; UV-B from 280 nm to 315 nm; and UV-A from 315 nm to 400 nm

operational for approximately 12 hours per day, and that the price of electricity is 8¢/kWh, the annual operating cost of a UV system is expected to be approximately $45, including the 35-watt UV lamp and the cost of electricity. It is assumed that the villagers provide their own storage tanks and sand filter; the raw materials for these components are readily available and inexpensive.

Based on the above assumptions regarding cost and system life-span, and a 12 percent discount rate, the total annualized cost of the UV system is approximately $100 per year. If the system operates for 12 hours per day, 8000 tonnes (8 million liters) of water is disinfected every year. Using a per capita drinking water requirement of 10 liters per day, a single system can provide enough water for approximately 2200 villagers. Accordingly, using a UV system to ensure potable water for a rural community of this size year-round costs about 5¢ per villager per year.

10. Integrated Resource Planning to Optimize Energy Services

In the foregoing example, an inexpensive, energy-efficient technology addresses a health problem of major concern in developing rural areas throughout the world, with the additional benefit of reducing the stress on supplies of traditional fuel supplies. Both the industrial and industrializing nations face the problem of supplying energy services at lowest cost, with available capital. Without energy efficiency, the annual bill for the developing countries will be $100 billion / year (according to the IEA reference in **Fig. 10**). This is twice the present level, and is nearly one-half of all investment in developing countries, which is unacceptable. With energy efficiency, many studies show that the $100 billion can be halved by 2010, thus the picture is much brighter.

During the first hundred years of public utilities, resources were constrained, and their mandate was to sell electricity and gas. The public paid little attention to the energy services (lighting, space and water heating, etc.) these carriers provided. With limits on available financing and environmental concerns, the role of utilities is changing: they should become sellers of "least-cost energy services," using "integrated resource planning" (IRP) as a tool. In the case of regulated, investor-owned utilities, it is necessary to rewrite the profit rules set by public utility commissions to make their lowest-revenue plan the profitable one.

IRP's advantages will vary by region. In the industrialized nations, ratepayers will save money through lower utility bills, and they will be able to afford better energy services. In the formerly planned economies and developing nations, customers will be able to buy more of the energy services they need for the same utility bills they pay now. There is too little capital available to grow energy supply at reasonable levels unless the efficiency of both supply and use improves drastically. Globally, efficient supply and use of energy will improve productivity and economic growth rates, and reduce the environmental threats associated with energy supply, including global warming.

Using IRP, utilities can identify and invest in the energy-saving technologies that pay for themselves through reduced energy bills the fastest. Examples of such technologies include purchasing controls that increase combustion efficiency, but also, purchasing the output of factories that manufacture compact fluorescent lamps, or energy-efficient appliances whose lifecycle cost is low but whose first cost is "too high" to customers with limited capital. Buying these products at wholesale prices and then leasing them to customers, keeps customer cash flow positive, meets utility obligations to serve its customers and maximize its profits for its shareholders.

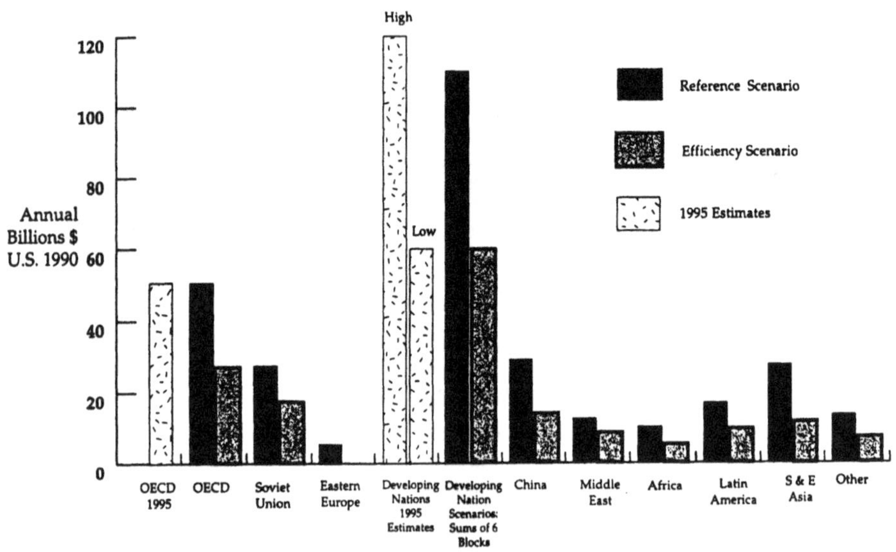

The OECD nations spend about $50 billion per year on its energy sector annually. Developing nations spend between $60 billion per year (based on USAID d and 120 billion per year (based on World Bank data). If the cash-strapped developing world's capital requirements are on the low side, then it may be able to meet its needs. However, if capital requirements grow to the $120 billion range, as even the USAID report suggests is likely, then developing nations will be unable to satisfy their growing energy needs. Thus, the energy-efficient path might be the only viable alternative.

Sources: "Energy Efficiency, Developing Nations and Eastern Europe," M.D. Levine et al., International Institute for Energy Conservation, 199 Steve Wiel; Mark Levine, LBNL, private communications, 1995. High Estimate by Steve Wiel using data from "The World Bank's Role in the Electric Power Sector," World Bank; low estimate by Mark Levine based on "Power Shortages in Developing Countries," USAID, 1988.

Figure 10.
Annual Energy Efficiency Capital Requirements, Actual and 1985-2025 Scenarios.

11. Conclusion

The many examples discussed in this paper show that it is simple and profitable to continue to improve energy efficiency. Every time an individual, family, business, factory or institution switches from consuming energy (supplied by imported fossil fuels) to investing in energy-efficient products (manufactured domestically), the local economy creates more local jobs, while a smaller number of jobs are lost in the Middle East or the North Sea. And it saves fossil fuel for future generations—powerful reasons to convince governments and utilities to accelerate their energy efficiency programs.

12. References

Additional references can be found at: http://EandE.LBL.gov/EAP/BEA/HIP/hiref.html

Akbari, H., S. Davis, S. Dorsano, J. Huang, and S. Winnett (editors). 1992. Cooling Our Communities: A Guidebook on Tree Planting and Light-Colored Surfacing. U. S. Environmental Protection Agency, Office of Policy Analysis, Climate Change Division.

Akbari, H., Rosenfeld, A., and Taha, H. 1989. "Recent Developments in Heat Island Studies, Technical and Policy," *Proceedings of the Workshop on Urban Heat Islands*, Berkeley CA, (February 23-24).

Akbari, H., Rosenfeld, A., and Taha, H. 1990. "Summer Heat Islands, Urban Trees, and White Surfaces," *Proceedings of American Society of Heating, Refrigeration, and Air Conditioning Engineers*, Atlanta, Georgia, (February); also Lawrence Berkeley Laboratory Report LBL-28308.

Alkari, V., Address to the Conference on the Use of UV Light for Water Disinfection, May 3, 1994, Bhubaneswar, Orissa, India.

Altman, L., "New Strain of Cholera Spreads in Thailand," *International Herald Tribune*, August 14-15, 1993, p.1.

Berdahl, P. and Bretz, S. 1994. "Spectral Solar Reflectance of Various Roof Materials," presented at the *Cool Building and Paving Materials Workshop*, Gaithersburg, Md., July 1994.

Hopkins, D. 1994 "Smart Thermal Skins for Vehicles,"27th International Symposium of Advanced Transportation Applications, Aachen, Germany, Oct 31-Nov 4, 1994.

Levine, M., A. Gadgil, S. Meyers, J. Sathaye, J. Stafurik, T. Wilbanks, 1991. "Energy Efficiency, Developing Nations, and Eastern Europe, A Report to the U.S. Working Group on Global Energy Efficiency." Washington D.C.: International Institute for Energy Conservation

Mills, E., 1994."A Graceful Exit," *International Association for Energy-Efficient Lighting Newsletter*, Vol. 3 No.8, pp 10-11.

Mills, E., 1995. "From the Lab to the Marketplace," U.S. Department of Energy, PUB-758.

Nadel, S., M. Shepard, S. Greenberg, G. Katz, A. de Almeida, 1991. Energy-Efficient Motor Systems. Washington D.C.: American Council for an Energy-Efficient Economy. pp 90-91

Rosenfeld, A., 1991. "The Role of Federal Research and Development in Advancing Energy Efficiency," Statement of Arthur Rosenfeld before James H. Scheuer, Chairman, Subcommittee on Environment, Committee on Science, Space and Technology, U.S. House of Representatives.

Rosenfeld, A., H. Akbari, S. Bretz, B. Fishman, D. Kurn, D. Sailor, H. Taha, 1995"Mitigation of Urban Heat Islands: Materials, Utility Programs, Updates," to appear in *Energy & Buildings*.

Sailor, D.J. and Kessler R. 1993. "Chapter 4.3: Three Dimensional Simulation Results," in *Analysis of Energy Efficiency and Air Quality, Interim Report,* pp. 4.57-4.82, Lawrence Berkeley Laboratory Report LBL-33051.

AN INTRODUCTION TO INTEGRATED RESOURCE PLANNING

A. T. DE ALMEIDA, P. FONSECA, N. SARAIVA
Department of Electrical Engineering
University of Coimbra
3030 Coimbra, Portugal

Abstract

The basic concepts and rationale of integrated resource planning (IRP) are explained. IRP methodology is characterised, and its pros and cons are discussed. The most promising technologies which can be used in IRP programmes are presented. The main factors governing the operation of IRP are explained.

1. Background

Electricity has been one of the essential ingredients to achieve technological development and to raise the standard of living of modern societies. Electricity consumption has been increasing exponentially for most of this century and the associated undesirable impacts of this growth became more visible. Those impacts include:

- Environmental impacts both at regional and at global level. These impacts can deteriorate the life-supporting capabilities of the environment. The operation of power plants, namely fossil fuel and nuclear plants, can affect the general quality of air, water and land in neighbouring regions or globally. Additionally those impacts may occur many years or decades after the beginning of the plants operation. The environment protection has become an important goal for the EU and all over the world, since a better environment is a key condition for improving the quality of life.

- Depletion of non-renewable energy resources. Although the world is not on the verge of exhausting those resources, the smooth transition to sustainable development requires a deceleration of the consumption of those resources.

- In some countries the expansion of the electricity supply system to cope with a fast growing demand lead a large financial debt of the utilities. This is due to the recent steep increase of capital costs to build new plants and also due to the regulation of

S. Nørgård et al. (eds.), Development with Sustainable Use of Electricity, 49-70.

electricity prices (in some cases this means prices well below marginal prices) for political reasons. In a growing number of developing countries the financial burden is also coupled with a capacity shortage of the supply system to meet the demand.

Traditionally electric utilities have essentially invested in supply-side options, although the most cost-effective combination of those options is chosen to meet the growing demand in a reliable way. The utilities have been applying sophisticated demand forecasting and planning methodologies to achieve those purposes, in most cases with successful results.

Electricity is a high quality and flexible form of energy which can be converted in basically all sorts of end-uses. However, consumers are not interested in consume kWhs, but they are interested in the energy services provided by the use of electricity.

The electricity is converted by a large variety of end-use equipment such as: motors (pumps, fans, compressors, etc.), lighting, etc., whose efficiency is normally quite far from the theoretical limits. Additionally, in plenty of situations, the investment in high efficiency equipment to reduce electricity consumption is more cost-effective than investing in new power plants. Cost-effectiveness is essentially dependent on the price of electricity and the relative costs of investing in power plant expansion and in conservation options. A combination of inertia and market imperfections has prevented the most cost-effective solutions from being systematically implemented.

Integrated Resource Planning (IRP) is a methodology which optimises electricity investments looking both to supply-side and to demand-side options to meet the growing demand for energy services. Additionally, externalities such as environmental impacts are also included in the evaluation process. This strategy besides bringing economic benefits, both to the utilities and to the consumers, also mitigates the magnitude of environmental damages. The use and depletion of primary energy resources is also decreased, therefore decreasing the dependency on imported energy resources.

Integrated resource planning has achieved substantial success in USA in terms of both electricity conservation, demand savings, and environmental and economic impacts. A comprehensive variety of conservation programmes were developed by USA utilities, which coupled with innovative regulatory frameworks are saving utilities and consumers billions of dollars. Due to its success, in most of USA integrated resource planning is either compulsory or is under consideration.

Integrated resource planning has not been widely implemented in Europe due the lack of a suitable regulatory framework, namely to be able to make a profit on large-scale demand side management projects. Additionally, there is a need to establish increasingly stringent minimum standards for air and water quality through legislation and regulation in the EU. The combination of profit regulation with environmental regulation can be effective to promote a significant implementation of IRP. USA experience may prove a valuable contribution for this purpose. For example, the EPA (Environmental Protection Agency) states that the pollution prevention, in a long term perspective, is less expensive than suffering environmental damages, mitigation adverse environmental impacts or imposing additional controls. Currently, 26 states of USA have requirements to consider externalities of electricity generation.

Integrated resource planning, if properly applied, will allow the utilities to adjust to a new operating environment and at the same time offers the possibility of more profitable operations. Utilities will sell more and more value added energy services instead of plain kWs and kWhs. It is thus likely that integrated resource planning may become an increasingly important, if not essential, methodology for most utilities.

2. Characterisation of Integrated Resource Planning (IRP)

2.1. WHAT IS INTEGRATED RESOURCE PLANNING

In conventional electric utility planning the strategy is to meet the forccasted electricity demand by optimising the mix of supply-side options (including power purchases), as well as the transmission and distribution systems, which can meet demand with a minimum cost. Additionally the planning process must also meet the specified reliability levels, and safety and environmental constraints.

Integrated resource planning (IRP), also called least-cost utility planning (LCUP) or least-cost planning (LCP), is a planning methodology with the aim of meeting in a reliable way the demand for energy services with minimum societal costs by optimising the mix of supply-side and demand-side options (see Table 1) taking into account externalities, such as environmental impacts.

In IRP the demand appears as a variable which can be influenced by the use of suitable demand-side options. Additionally, demand-side options are considered on an equal level with supply-side options not only in terms of kWh costs, but also in terms of reliability, controllability and external impacts. Because traditional planning has neglected demand-side measures, the introduction of integrated resource planning may produce more emphasis to conservation measures.

Table 2 shows a comparison between traditional utility planning and integrated resource planning. The term "least-cost planning" to describe the strategy to provide least-cost energy services was popularised by Roger Sant (Sant, 1979). In USA the use of the term integrated resource planning (IRP), instead of least-cost planning, is preferred since cost minimisation was also a goal of conventional planning approaches. Thus IRP is more appropriate and accurate term for describing the process of optimising the balance mix of supply and demand options to meet the demand for energy services.

TABLE 1. Examples of resource options (adapted from NARUC, 1988)

Supply-side options	**Demand-side options**
Conventional plants: Large fossil-fueled Nuclear Combined-cycle gas turbines Small combustion turbines Life extensions of power plants Transmission and distribution expansion	Energy-efficiency options (customer): Home weatherization Energy efficient appliances Efficient lighting Passive solar modifications High efficiency heating, ventilating and A/C Efficient motor systems
Non-utility-owned generation: Small-scale hydro Self-generation Independent power producers Cogeneration	Energy-efficiency options (utility): Reduced transmission and distribution losses Advanced transformers Load management Control of appliances
Purchases: Requirements transactions Coordination transactions	Rates: Time-of-use Incentive Interruptible
Renewables: Geothermal Solar Wind	

TABLE 2. Differences between traditional planning and integrated resource planning (Hirst, 1991)

Traditional planning	**Integrated resource planning**
Focus on utility-owned central-station power plants	Diversity of resources, including utility-owned plants, purchases from other organisations, conservation and load-management programs, transmission and distribution improvements, and pricing
Planning internal to utility, primarily in system and financial planning departments	Planning spread among several departments within utility and often involves customers, public utility commission staff, and nonutility energy experts
All resources owned by utility	Some resources owned by other utilities, by small power producers, by independent power producers, and by customers
Resources selected primarily to minimise electricity prices and maintain system reliability	Diverse resource-selection criteria, including electricity prices, revenue requirements, energy-service costs, utility financial condition, risk reduction, fuel and technology diversity, environmental quality, and economic development

2.2. WHY THERE IS A NEED FOR INTEGRATED RESOURCE PLANNING

In a perfect market economy with proper price signals, the consumers would make the proper decisions in choosing the least-cost demand-side options. Therefore the critics of IRP point out that if IRP is really cost-effective, consumers should need no help from the utility in making cost-effective investments in conservation measures. However a combination of consumer inertia and market imperfections have prevented a stronger penetration of conservation options. The main barriers include: (a more detailed description can be found in the chapter "Energy Efficient Motor and Drives")

- The information gap
- The payback gap
- Capital availability
- Split incentives
- Utility rates
- Uncertainty over future rates
- High equipment costs

For all these reasons there is a substantial efficiency gap between the energy efficiency of the average new investment investments and efficiency of the most cost-effective options. IRP has the capabilities and appears to be the right strategy to overcome the above barriers.

2.3. INTEGRATED RESOURCE PLANNING METHODOLOGY

Integrated resource planning methodology can be decomposed in a number of discrete steps (adapted from NARUC, 1988):

- Identifying the objectives of the plan (e.g., reasonable price for consumers, reliable service, minimal environmental effects, economic development, flexibility to cope with risk and uncertainty).

- Developing load forecasts (normally a base, high and low forecast) for the different sectors, corresponding to different assumptions about economic growth, demographics and energy prices.

- Determining the levels of capacity expected for each year of the plan.

- Identifying available options (e.g., options shown ir able 1) required to bridge the gap between expected loads and capacities.

- Evaluating all the resources in a consistent fashion in terms of costs, reliability, controllability, environmental impacts, etc. An important phase of this step, not found in traditional planning, is the determination of the conservation cost-supply curves. These curves relate the cost (in terms of ECU/kWh and ECU/kW) of conserving electricity and peak power and the associated savings potential for each demand-side option.

- Selecting the most promising options to produce a cost-effective, flexible and responsive plan.

- Integrating methods of supply-side options with demand-side options.

- Constructing scenarios, pitting the selected mixes of options against possible economic, environmental, and social circumstances.

- Evaluating the economic and technical success of each mix of options under the circumstances of the various scenarios.

- Analysing the uncertainties associated with each possible plan of action.

- Screening the alternatives to eliminate those that are not suitable.

- Rank ordering of the alternative courses of action.

- Testing each alternative for cost-effectiveness from a variety of viewpoints (e.g., the utility, ratepayers of different classes, and society).

- Re-evaluating the alternatives considering economic, environmental, and societal costs.

- Selecting a plan for implementation, one that most nearly satisfies all the objectives of the plan.

- Developing a plan of action.

- Implementing the plan of action to bring about the least-cost provision of electric power.

- Monitoring and evaluating the operation of the utility under the plan and revising if necessary.

2.4. CHARACTERISTICS OF A GOOD LEAST-COST PLAN

In many cases the utilities prepare least-cost plans to comply with requirements from the regulatory commissions. The plan is intended to inform all stakeholders (utility, consumers, regulatory commissions, government authorities, environmental groups, etc.) of the alternatives to meet the demand for energy services. The plan also points the preferred path and the respective action plan to achieve it.

A good least-cost plan should meet the following criteria (Hirst 1990):

2.4.1. Clarity of plan

Adequately inform various groups about future electricity resource needs, resource alternatives, and the utility's preferred strategy. The plan should feature:

- Clear writing style
- Comprehensible to different groups
- Presentation of critical issues facing utility, its preferred plan, the basis for its selection, and key decisions to be made.

2.4.2. Technical competence of plan

It affects utility decisions on, and regulatory approval of, resource acquisitions. The plan should include:

- Comprehensive and multiple load forecasts
- Thorough consideration of demand-side options and programs
- Thorough consideration of supply options
- Consistent integration of demand and supply options
- Thoughtful uncertainty analyses
- Full explanation of preferred plan and its close competitors
- Use of appropriate time horizons

2.4.3. Adequacy of short-term action plan

Provide enough information to document utility's commitment to acquire resources in long-term plan, and to collect and analyse additional data to improve planning process.

2.4.4. Fairness of plan

Provide information so that different interests can assess the plan from their own perspectives. They should be made with:

- Adequate participation in plan development and review by various stakeholders
- Sufficient detail in report on effects of different plans

3. Pros and cons of integrated resource planning (IRP)

3.1. ADVANTAGES OF IRP

The introduction and application of IRP methodologies by electric utilities can bring a wide range of advantages. Similar benefits can be obtained if IRP is applied in other fields such as gas utilities. The potential benefits of IRP include:

3.1.1. Greater economic efficiency
The mix of supply and demand side options provided by IRP provides energy services with minimum costs, leading to a decrease of the cost of manufactured goods and services. Consumers will pay less for their electricity bills and thus have more disposable income for other purposes.

3.1.2. More flexibility, less risk and the possibility of more profitable operations for the utilities
IRP is a win-win situation for consumers and utilities. The need to invest in large power plants, with the associated financial burden and risks, will be reduced. Cost-effective demand options provide a more flexible and shorter term approach to follow demand. Additionally the value of energy savings provided by conservation investments keeps up at least with inflation. The possibility of the utility selling value added energy services can also mean more profit opportunities for the utilities.

3.1.3. Substantial reduction of environmental impacts
In terms of the public opinion in developed countries, this seems the strongest argument in favour of the application of IRP. IRP leads to a reduction of electricity consumption which means that less generation is required, which in its turn leads to a reduction in the emission of pollutants associated with the operation of power plants. Emission of pollutants such as CO_2, NO_x, SO_2 and radioactive materials, and their associated effects in the environment can thus be mitigated. Although the emission of certain types of pollutants, such as SO_2 and NO_x can be controlled at a cost, there are no technologies available to reduce CO_2 emissions once it is released. Electricity conservation appears as one of the most important strategies to reduce CO_2 emissions, and thus help to reduce the threat of global warming.

3.1.4. IRP takes a more balanced perspective in terms of societal benefits
The minimisation of the costs of energy services also takes into account externalities such energy security, economic impacts and environmental impacts (EPRI, 1991), which are increasingly relevant for modern society.

3.1.5. IRP can help to improve the relations between the utility and the consumers
To prepare and implement a good IRP plan the utilities must know well the consumer needs. Consumer groups can also participate in the IRP process. This improved

knowledge, as well as the savings provided by IRP, can contribute to improve relations and to provide a better public image of the utilities.

3.1.6. IRP can improve the already good load forecasting

The detailed knowledge of the disaggregated electricity consumption of the different sectors, required for the IRP process, can improve the accuracy of load forecasting. This further contributes to a reduction of the investments costs, as there is less uncertainty when planning new investments.

3.1.7. Reduction of the consumption of non-renewable energy resources used in power plants

This reduction is a positive contribution to a smooth transition to achieve sustainable development in a not too distant future.

3.1.8. Reduction of dependency on imported fossil fuels (gas, coal and oil) leading to greater energy security and to an improvement in the trade balance.

3.1.9. Development of new business and employment opportunities

Large-scale implementation of electricity conservation measures will promote the appearance on the market and the penetration of high-efficiency equipment. Energy conservation measures have also been shown to contribute in a positive way to increase employment (Nadel, 1991b), not only in activities related to equipment manufacture but also in auditing, installation and evaluation.

3.1.10. Example for developing countries to follow

Although energy and electricity consumption per capita stands at very modest level in developing countries, it has been shown that there is a substantial cost-effective savings potential. The success of IRP in developed countries, measured in terms of economic efficiency and emissions per capita, can exert a large influence in the widespread adoption of IRP in those countries.

3.2. DISADVANTAGES OF IRP

IRP may also have some undesirable effects which include:

3.2.1. Higher rates

Electricity rates may need to go up in order that the utility can finance conservation measures. The need for this increase will depend upon the relation of electricity prices with marginal costs and upon the degree of financing given to the conservation measure. In a properly implemented IRP action, even if the rates go up the kWh consumption should go down by a greater amount, leading for most consumers to a reduced electricity bill.

3.2.2. Equity problems
As some consumers may see their electricity bill increased, while for others is decreased. If electricity rates go up, consumers which either installed conservation measures before the IRP plan is implemented or non-participating consumers will pay more for the same level of consumption.

3.2.3. Possibility of the utility being forced to make charity
Non-participating consumers are mostly low-income groups. Thus, pressure may be placed on utilities to give extra support, which may mean full payment of the conservation measures, to those groups. This type of financial help, if required, should not be forced on the utility, but assumed by social welfare organisations or governmental organisations.

3.2.4. Negawatts (energy savings) are more difficult to meter than megawatts Measuring energy savings are particularly relevant in situations when a utility is financially rewarded (or penalised) as a function of the results of the performance of the conservation investments.

3.2.5. Variable DSM costs
Some of the costs associated with conservation investments vary from consumer to consumer and are difficult to quantify. These costs include consumers time, equipment down-time, inconvenience and discomfort incurred by consumers when learning about, evaluating, installing, adjusting to, re-evaluating, etc., demand-side options (Ruff, 1987). These variable costs can lead to considerable uncertainty on the cost and savings potential of demand-side options.

3.2.6. Uncertainty in the achieved savings
A gross error in these savings is equivalent to capacity shortfall in the supply-side. The achieved savings depend not only on the cost-effective savings potential but also on the penetration rate of the conservation measures. There is considerable uncertainty on the estimation of the public acceptance, although there are means to improve the acceptance rate, namely the use of incentives and regulation.

3.2.7. A slowdown in activities related to manufacturing and installation of generation, transmission and distribution equipment
This is a logic consequence of the decrease of the rate of expansion of utility system. However experience has shown there is a net increase in employment, specially at local level, when there is a large scale implementation of conservation measures. (Nadel, 1991b).

3.2.8. Retraining of staff
Utilities have a large number of staff whose expertise has been focused in delivering electricity to consumers. The implementation of IRP and associated conservation

measures requires expertise in converting electricity in energy services. This requires shifting of staff activities with the need for a substantial retraining effort. Another possibility is the use of Energy Service Companies (ESCOs) for the evaluation and installation of the conservation measures in the customer premises, although with this strategy the utility looses some potential business.

4. Factors governing the operation of IRP

4.1. MEASURES AND INSTRUMENTS OF ENERGY POLICY

In order to improve electricity efficiency, the most important types of measures are:

- Economic instruments (prices, taxation and financial incentives)
- Information and education
- Normative measures (labelling and standards)
- Research, development and demonstration.

These measures can be implemented by energy authorities, utilities, and by the manufacturers together with research laboratories. The energy authorities can have the responsibility in all the above measures, although their main role is in the definition of prices, taxation and normative measures. Research laboratories and manufacturers (with the possible support from energy authorities) play a key role in research, development, demonstration and market development of new high-efficiency technologies.

Utilities can also have a decisive role in removing the market and institutional barriers which prevent the penetration of cost-effective high-efficiency equipment (Nadel, 1991b). In order to increase the penetration of conservation measures the utilities can use several methods, namely :

- Information and training
- Audits
- Zero interest loans
- Rebates
- Competitive bidding
- Direct installation by the utility

Information measures are directed at mitigating the consumer information gap. Although inexpensive, to be effective information needs to be coupled with other measures, such as audits and incentives. *Training* is normally used to provide information in energy management expertise to operators of commercial and industrial facilities.

Free audits, or at low cost, take information a step further, by helping customers to identify and rank energy conservation opportunities. Audits alone produce low electricity savings.

Zero interest loans, can be used to overcome the barrier related to lack of capital for financing the conservation investment.

Rebates are introduced to overcome the payback gap, that is to reduce consumer payback to acceptable values. Normally the penetration rate increase with the relative value of the rebate, with a steep increase in the case of large rebates. Marketing strategies and technical support services also have a decisive importance on programme participation and savings.

In competitive bidding programmes (see section 4.6) the implementation of the conservation measures (auditing, installation and follow-up) are normally carried out by an energy service company, which is rewarded as a function of the results achieved. This can lead to high penetration and savings.

Direct installation programmes by the utility can achieve higher penetration and electricity savings, although it is more expensive for the utility than the information, audit, loan and rebate programmes. This type of programmes is specially appropriated for hard-to-reach customers (e.g. low-income customers) and for utilities having a capacity short-fall in the near term.

4.2. PRICES

The tariffs are the most important tool to influence supply and demand. Behavioural studies show that most consumers will be influenced in the choice of high-efficiency equipment or will change their demand in response to price signals, although tariffs alone are not enough to ensure the penetration of high-efficiency equipment (IEA, 1989).

In most European countries the rates are based in historical average costs, which are normally lower than avoided costs. In order to achieve economic efficiency in selection and operation of end-use equipment, the rates should reflect the long-term marginal costs (both for energy and peak power) of the system, and prices should also reflect external costs. To ensure this purpose the regulatory authorities play an important role. Ideally, each customer should pay the respective costs in order to avoid cross-subsidisation. The price structure should also be stable and be able to motivate the consumers to take the most cost-effective decisions.

Recent technological developments have a large potential to achieve efficient pricing. Modern electronic meters, which can communicate with the utility, have the capability of tracking the real-time price of electricity, thus providing the right signals for the consumers to make proper demand-side investments. Several field trials, involving different technologies, are under way in several European countries (Finland, Holland, Italy, Spain and UK) and in the USA, involving several thousand of customers, mainly in the residential sector. The pioneering experiments with smart meters were started in the mid eighties in UK, and a large field trial of 100,000 consumers is under consideration in that country.

All categories of consumers deserve equal treatment and economic efficiency in the tariffs. Specially important is to differentiate the price of electricity to consumers which require different levels of reliability. Reliability is a variable increasingly considered to be determined by economic considerations rather than by engineering standards. Modern communication, control and metering technologies can be used for price-reliability differentiation to decrease peak demand and to save valuable reserve margin.

4.3. TAXATION

By influencing prices, taxes are also a powerful tool to direct the market forces in the desired direction. Electricity is taxed in all countries, but in general it receives a more favourable treatment than other forms of energy, specially in comparison with oil products. Additionally, the electric utilities normally also benefit from tax exemptions or reductions. These factors lead to a further undervaluing of electricity, thus creating market distortions between electricity and other forms of energy. This is particularly relevant in areas where there is competition between electricity and other fuels, such as the space heating market. The main purpose of the tax has been to raise revenue for financial reasons, but taxes may also be used to reflect societal costs such as:

- Environmental impacts (see section 4.5), such as global warming, acid rain, health effects, etc.
- Depletion of non-renewable resources
- Economic impacts (e.g. energy security, balance of trade)

A programme to be taken into account when legislating about energy taxation may be that of the Danish Government: the Energy 21 plan. In fact, this package of laws aims two main goals: to decrease significantly the level of pollutant emissions (20% of CO_2; 80% of SO_2 and 30% of NO_x) by the year 2005 and to promote independence in relation to fossil fuels.

So, in order to achieve these goals, a substantial taxation package has been approved, mainly consisting on:

TABLE 3. The New Danish Energy Tax System

Heavy processes:	A CO_2 tax of DKK 5 per tonne CO_2 in 1996 to be stepped up gradually to DKK 25 per tonne CO_2 in 2000. Reimbursements are negotiable, effectively reducing the tax rate to DKK 3, provided enterprises accept to implement an energy-saving action plan.
Light processes:	The current CO_2 tax to be stepped up gradually from DKK 50 in 1996 to DKK 90 per tonne CO_2 in 2000. Enterprises with a high level of energy consumption can opt for an agreement obliging them to undertake an energy saving action plan, even though the relevant production process is not considered to be heavy. In this event the enterprise will receive a reimbursement.
Space heating:	Energy and CO_2 taxes to be stepped up gradually to the same level of taxation as applies to households, i.e. an average level of approx. DKK 600 per tonne CO_2.
A tax is to be introduced gradually on all energy consumption at a rate of DKK 10 per kg SO_2 with special reimbursement schemes for the consumption of coal and fuel oil for certain enterprises.	

Furthermore, this programme foresees not only the reimbursement of the enterprises which agree to invest in energy efficient programmes but also a full tax liability to those who decide not to do so. Another possible strategy, an universal energy tax, introduced unilaterally and without compensatory measures, would be detrimental to the competitiveness of a huge number of enterprises, increasing unemployment unacceptably. To avoid this, it has been decided that all the revenue raised by the energy tax should be recycled to industry and that the rates of taxation shall be differentiated according to the level of energy consumption in the various production processes.

4.4. LABELLING AND STANDARDS

Labelling provides the means to transmit to the consumer the energy performance of end-use equipment. Most consumers, specially in the residential sector, are not aware about the efficiency and energy costs of the equipment. The purchase decision is normally influenced by other factors (price, aesthetics, convenience, etc.). To be effective labelling should state the equipment efficiency, the operating costs under typical operating conditions (e.g. number of operating hours, electricity prices) and the relative performance of the equipment compared to models from other manufacturers.

Labelling, if compulsory, can help to bridge the information gap which is one of the factors affecting the poor choice of equipment. Additionally, labelling increases competition between the different manufacturers to improve the efficiency, specially if it provides comparative performance among manufacturers.

Efficiency standards go a step further than labelling, by defining minimum efficiency levels of new equipment and buildings. Building codes and equipment efficiency standards can achieve savings beyond those achieved by utility programmes. For example, the California Energy Commission estimates that in 1983, building codes and appliance efficiency standards were responsible for 53% of the 2700 MW of the savings in the peak demand, compared with 45 % due to utility programmes (Nadel, 1991b).

At little administrative cost, standards are effective to overcome barriers such as:

- Lack of information on the energy performance and associated costs of the equipment.
- Even if the consumer is aware of the energy performance of the equipment, the relative low operating costs of some appliances, (specially in the case of home appliances) may have little influence in the consumer's decision.
- Market inertia to introduce high-efficiency equipment (this is the case of high-efficiency motors, as the market is dominated by sales to original equipment manufacturers)
- Split incentives barrier (the buyer of the equipment is not going to pay the operating costs).

Standards also stimulate innovation and competition among manufacturers, besides removing inefficient equipment from the market. The setting of efficiency standards should be based on realistic and cost-effective goals, defined in cooperation with manufacturers and research laboratories. Standards also prevent low-efficiency and low price imports from entering the market.

In the evaluation of conservation measures, some end-use metering could be avoided, as well as their associated costs, if appliances had their power consumption and efficiency measured by independent test laboratories. Efficiency measurement standards can differ significantly in different parts of the world (e.g. electric motors and refrigerators). Additionally, some manufacturers place nameplate data which sometimes exhibits a large difference in relation to the real value. This holds particularly true for high penetration appliances such as compact fluorescent lamps. Moreover other relevant characteristics such as lifetime should be measured. The independent testing information should be passed to consumers, utilities and consumer groups.

4.5. ENVIRONMENTAL EXTERNALITIES

An externality is a residual or side effect of an economic activity in which a benefit or cost is conferred upon a third party (a party not involved in the original transaction either as a producer, consumer, or agent); e.g. air pollution from a coal-fired power plant (Ottinger, 1990).

In the last decades, environmental impacts have become an important issue in electric utility planning. In USA, a true IRP should account for all the costs, including any environmental costs or benefits, associated with the options considered in the plan. All

environmental effects associated with the different electricity generation options should be quoted in IRP if it is a really least cost one.
Associated with electric power generation, there are important environmental damages, which depend on the type of the resource used in the process of producing electricity. Some of those environmental damages are air pollutants (SO_2, No_x, etc.), greenhouse gas emissions (CO_2, etc.), quality of water and land, etc.

Figure 1 shows the combined carbon dioxide (CO_2) and methane (CH_4) emissions expressed as tons of CO_2 equivalent per million kWh of electricity produced from the operation of five electric operating technologies.

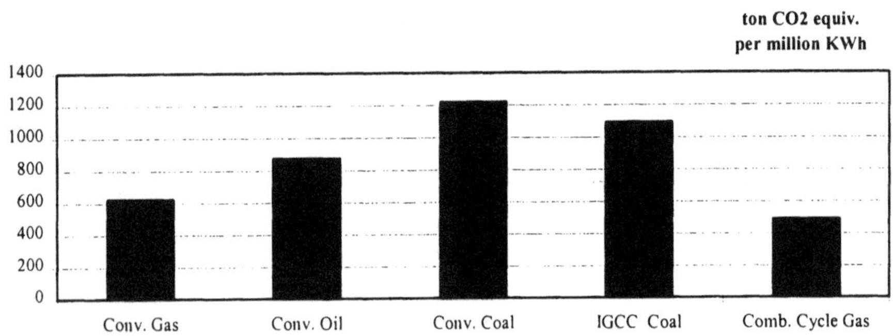

Figure 1.
CO_2 and CH_4 Emissions from Electricity Technology Operations, assuming one mole of CH_4 has the radiative effect of 25 moles of CO_2 (a ratio of nearly 70 on a weight basis).

As it can be seen, combined cycle gas produce less greenhouse gas emissions than the other operating technologies. It produces approximately half of the emissions produced from the operation of a conventional coal power plant, which represents one of the main benefit of combined cycle gas.

Environmental externality costs of electricity are costs to society (human health, property damage and other environmental damages) resulting from electricity production, whose costs are not already incorporated in the price of electric services.

There are different methods for calculating environmental externality costs. Variables like fuel switching costs, pollution control costs, population exposed, discount rates, and others, are some factors that affect environmental costs. In the evaluation of the environmental costs two methods have been used:

- Calculating the environmental damage costs associated with the impacts of each phase of the fuel cycle of a given generating technology. The extent of each impact is estimated and a value calculated by the sum of the partial impacts. The main difficulty of this method is the quantification of the monetary value of each impact.

- The value of pollution reduction is based on the cost of controlling the pollutants emitted by the generating technology. This method, although simple, is not suitable for CO_2 costing and the value of controlling emissions is not related to the societal costs.

The incorporation of environmental externalities in the IRP process can strongly influence the balance of demand and supply options chosen, the timing of the additions and the operating strategies of present and future resources.

The inclusion of quantitative environmental impacts was required for the first time in 1980, in the Pacific Northwest Electric Power Planning and Conservation Act. This act required environmental costs and benefits to be explicitly and rigorously evaluated in the evaluation of the alternative options to expand the system.

Several approaches have been proposed to internalise environmental costs, namely:

- Regulation
- Corrective taxes
- Tradeable limits

Regulation normally takes the form of standards, either on a technology or performance basis. Technology standards may specify for example the use of "Best Available Control Technology", to limit the emissions of a pollutant such as SO_2. Performance based standards specify emission levels and leave the choice of technology to the user. However, standards and regulations only limit emissions to established levels and do not eliminate pollution. The environmental impacts that remain after regulation are significant and are ignored.

Corrective Taxes allow the internalisation of the societal cost of pollution in the total cost, in the following way:

$$Tax = Societal\ Cost - Supply\ Production\ Cost \qquad (1)$$

Ideally, regulation and taxes should lead to similar results. When investing in pollution control, the socially efficient solution is achieved, when the marginal cost of pollution control is equal to marginal societal cost of pollution.

In the cases where there is no pollution control technology available, as is the case of CO_2, the use of carbon taxes seems appropriate to internalise the environmental costs.

Tradeable permits allow one utility to sell or to buy emissions quotas to/from other utilities, thus creating a competitive market in environmental emissions. This is a flexible approach which encourages the development of improved emission control technologies and can lower the overall cost of achieving a certain level of emission reduction.

Although there has been considerable evolution on the scientific understanding of utilities' environmental impacts, there is still considerable uncertainty on the quantitative assessment of those impacts. Valuing the risks of nuclear proliferation and the psychological effects of living near a nuclear plant or coal facility are important areas where research is needed. These areas are important factors when assessing the environmental costs.

Due to the uncertainties in the estimating the costs of environmental impacts, the estimated costs in the IRP process vary from one region to another. A few examples in USA are (EPRI 1991):

- In New York the environmental costs are calculated through a scoring matrix to take into account different externalities. These costs can be up to 1.4 cents /kWh or up to 24% of the avoided cost.

- In Wisconsin there is an automatic credit of 15% for non-combustion resources.

- In Vermont there is 15% deduction of the costs for demand-side measures (5% for environmental benefits and 10% for risk reduction).

- In California a surcharge of 10-30% on generation costs is being proposed.

4.6. COMPETITIVE RESOURCE BIDDING

In order to increase the energy efficiency of the supply system the 1979 Public Utilities Regulatory Policies Act (PURPA) was issued (Stoll, 1989). This piece of legislation requires that the utility buys the electricity generated by independent power producers (cogeneration and small power producers) at the utility's avoided cost. In some states long term contracts based on the projected avoided costs were also offered.

PURPA increased substantially the amount of non-utility generating capacity addition, from 500 MW/year in the mid-seventies to 3000 MW/year in the late eighties. PURPA marked the large-scale introduction of nonutility operators in the electricity market.

After the first oil energy crisis with the rise in energy prices, another important type of operators appeared in the energy scene. Energy Service Companies (ESCOs) were quick to appear the scene to take advantage of very cost-effective opportunities for energy conservation. ESCOs started working mainly with large industrial and commercial consumers using third-party financing schemes. ESCOs built substantial experience and can bridge the lack of utility skills in evaluating and delivering demand-side options which are cost-effective for the different types of consumers.

During recent years several USA utilities have carried out competitive integrated resource bidding. In this process independent producers, electricity distributors, large industrial and commercial customers, and third parties (e.g. ESCOs) bid to supply a certain amount of capacity to an utility, either through supply-side or demand-side options. The bids are then ranked based on societal cost-effectiveness. The utility's scoring system should explain in a clear way the assessment criteria and the weights for the different features.

The selected demand-side bids normally involve the participation of ESCOs and customers, in a contractual agreement with the utility that specifies the amount of savings over defined time frame, on performance contract basis. The successful bidders usually provide a package of services to the utility's customers that include engineering analysis of their facilities, acquisition and construction management of energy efficiency improvements, operation and maintenance of equipment over 5 to 10 years, and financing. The amount of selected demand-side options is normally between 5 and 20% of the utility's need for new resources and in the range 10-50 MW (Hirst, 1991). For utilities, competitive resource bidding may provide excellent business opportunities if the utilities create their own ESCOs to exploit the market.

4.7. INCENTIVE MECHANISMS FOR THE UTILITIES TO CARRY OUT IRP

Traditionally the utilities make a profit on sales and had an incentive to increase the sales to achieve higher profits. The implementation of large-scale demand-side programmes requires decoupling profits from sales. For this purpose, several states in USA have implemented the *Electric Revenue Adjustment Mechanism (ERAM).*

ERAM is a rate adjustment mechanism that accounts for the over- or undercollection of revenues that occurs when actual electric sales differ from sales forecasts used to set the rates. ERAM eliminates impacts on utility earnings due to sales fluctuations resulting from conservation and load management programmes, weather, and other forecast conditions.

Most regulatory commissions allowed utilities to consider investments in conservation measures as operating expenses and thus suitable to be deducted from current revenues. This treatment of demand-side investments, called expensing, was the most common during the mid and late eighties. Under expensing the utilities can recover conservation investments but do not make a return on each unit of electricity saved. However, more recent approaches provide more incentives to utilities than expensing. The profits can be linked to the demand-side investments by two mechanisms (CEC, 1990):

- *Rate of return*, also called rate basing, which creates demand-side assets that are eligible to earn a rate-of-return like traditional supply-side investments. This approach is sound when the investment is socially optimal, but the marketplace does not provide a fair return to the utility due to market imperfections.

- *Shared savings*, defined as a percentage of the difference between the costs of conservation and load management programmes and the costs of avoided supply evaluated over the life of the resource. The higher the net savings of the conservation measure, the higher the utility's profit. Although recently introduced, the shared savings approach is gaining wide acceptance due to its market-oriented approach.

On the other way the utilities must also agree to penalties, when the performance results of the demand-side investments do not meet the required minimum levels. Both for the determination of the shared savings and of the performance penalties, there is a need for measuring and evaluation of the conservation programmes.

4.8. PROCEDURES FOR THE EVALUATION OF THE COST-EFFECTIVENESS OF CONSERVATION MEASURES

A number of tests is being used in USA for calculating the cost-effectiveness of conservation measures (CEC, 1990):

- *Participant Test*: Measures the difference between the quantifiable costs incurred by a participant in a demand-side programme and the subsequent cost savings received by that participant. From this perspective, a programme is cost-effective if the value to the participant exceeds the costs borne by the participant.

- *Ratepayer Impact (RIM) Test*, also called the *No-Loosers Test*: Measure of the distribution equity impacts of conservation programmes on nonparticipating utility ratepayers. From this perspective, a programme is cost-effective if it reduces a utility's rates. This test is mainly used to evaluate the rate impact and it is not normally used as a rejection criteria.

- *Utility Cost Test*: Measures the difference between the costs incurred by a utility (programme costs) and the avoided supply costs (fuel operating and capacity costs) due to a demand-side programme. Costs and benefits incurred by programme participants are excluded from this test. From this perspective, a programme is cost-effective if the avoided supply costs exceed programme costs.
- *Total Resource Cost (TRC) Test*: Measures the difference between the total costs of a demand-side programme (including costs incurred by the utility and the participant) and the avoided costs of utility supply.

From this perspective, a programme is cost-effective if the avoided supply costs exceed the total programme costs.

- *Societal Test*: A variant of the TRC test which includes quantified effects of externalities (such as environmental costs) in the measures of costs and benefits.

The application of these tests allows the evaluation of the impact of demand-side measures on the different stake-holders of the planning process. For a fair evaluation the supply-side measures should be judged using the same tests as the demand-side options. Although the societal test gives a more global perspective, the final choice and tuning of the programmes will normally also take into account the results of the other tests.

Higher rates are inequitable to consumers which do not participate in a given programme and thus pay higher bills. The solution to this problem is to offer all consumers a variety of attractive programmes to minimise the number of non-participants. If all the consumers take at least a few of the options available to them, even if a particular programme has a few nonparticipants, almost all consumers will pay lower bills.

5. Bibliography

[1] CEC 1990, "*An Energy Efficiency Blueprint for California - A Report of thc Statewide Collaborative Process*", California Energy Commission (CEC), January 1990.

[2] De Almeida, A., Greenberg, S. and Blumstein, C., 1990, "*Demand- Side Opportunities Through the Use of Energy-Efficient Motor Systems*", IEEE Transactions on Power Apparatus and Systems, August 1990.

[3] De Almeida 1991, "*Technology Overview*" , De Almeida, A., International Conference on Demand-Side Management and Least Cost Planning", Copenhagen, October 23-24, 1991.

[4] De Almeida, A., Rosenfeld, A, Roturier, J. and Norgard, J., "*Integrated Electricity Resource Planning*", Kluwer Academic Publishers, Boston, 1994.

[5] EPRI 1991, "*Environmental Externalities: An Overview of Theory and Practice*", EPRI report CU/EN 7294, Electric Power Research Institute, Palo Alto, USA, May 1991.

[6] Hamrin, J., Rader, N., (1993), "*Investing in the Futur - A Regulator's Guide to Renewables*", The National Association of Regulatory Utility Commissioners, Washington, February 1993.

[7] Hirst 1990, "*Measuring Performance: Key to Successful Utility Demand-Side Management Programs*", Hirst, E., Oakridge National Laboratory, May 1990

[8] Hirst, E. and Goldman, C., 1991, "*Creating the Future:Integrated Resource Planning for Electric Utilities*", Annual Review of Energy, Volume 16, 1991

[9] IEA, 1989, "*Electricity End-Use Efficiency*", International Energy Agency, Paris, 1989

[10] ISBN 87-7844-062-9, (1996), "*Energy 21 - The Danish Government's Action Plan for Energy*", Copenhagen, June 1996.

[11] ISBN 87-7856-030-6, "*Energy Tax on Industry in Denmark*", Copenhagen, December 1995.

[12] NARUC 1988, "*Least-Cost Utility Planning Handbook - A Handbook for Public Utility Commissioners*", National Association of Regulatory Utility commissioners, Washington, USA, 1988

[13] NARUC 1994, "*Fifth National Conference on Integrated Resource Planning*", National Association of Regulatory Utility Commissioners, Kalispell, Montana, USA, 1994

[14] Nadel , S., 1991b, "*Electric Utility Conservation Programmes: A Review of the Lessons Taught by a Decade of Program Experience*" in "State of the Art of Energy Efficiency", edited by Vine, E. and Crawley, D., American Council for an Energy-Efficient Economy, Washington, 1991

[15] Nadel, S., Sheppard, Greenberg, S., Katz, G. and De Almeida, A., 1991, "*Energy-Efficient Motor Systems*", American Council for an Energy- Efficient Economy, Washington, 1991.

[16] Ottinger, R., L., and al, (1990), "*Environmental Costs of Electricity*", Oceana Publications, Inc, New York.

[17] Ruff, L., 1987, "*The Basic Economics of Utility Least-Cost Planning and Demand-Side Management*", Northeast Utilities, October 1987

[18] Sant, R., 1979,"*The Least-Cost Energy Strategy*", Carnegie-Mellon University Press, Pittsburgh, USA, 1979.

POLISH EFFICIENT LIGHTING PROJECT-THE DSM PILOT EXPERIMENT

ADAM GULA and ZBIGNIEW HANZELKA
University of Mining and Metallurgy and
The Polish Foundation for Energy Efficiency
Kraków, Poland

MARC R LEDBETTER and ROBERT G. PRATT
Pacific Northwest National Laboratory
Portland, Oregon, USA

Abstract

A Demand Side Management (DSM) pilot project of electricity saving by a wide installation of compact fluorescent lamps (CFLs) carried out in the city of Chelmno in Northern Poland is described. The initial results of the measurements show a significant (ca. 15%) power peak reduction at the 0.4 kV level. No significant increase in the harmonics distortions or neutral wire current are observed within the measurement accuracy. Finally, the DSM promotion sales system has led to an unexpected growth in customer interest in the efficient lamps and in a high saturation of the project-target areas with CFLs (about 5 CFLs per household).

1. Introduction

A characteristic feature of the Polish power sector is the 90% domination of hard and brown coal fuels [6]. The use of these fuels is accompanied by the emission of greenhouse gases and, despite installation of pollution control equipment which is continuously being upgraded, there continue to be significant emissions of particulate matter and gases causing acid rain. Besides the airborne pollutants, rapidly growing ash dumps are also a significant pollution factor. An efficient way to reduce those types of environmental damage is the reduction of energy consumption, particularly electricity. The analysis of the energy consumption structure in Poland indicates that the domestic share in the total energy consumption accounts for 30-40% [6]. This is a very high value in comparison with highly developed countries, and holds the potential for hitherto only partially exploited energy savings.

Among other factors, consideration of these facts has given rise to the idea of reducing domestic electricity consumption by replacing traditional filament-bulb

S. Nørgård et al. (eds.), Development with Sustainable Use of Electricity, 71-96.

lighting with modern energy-efficient systems - the compact fluorescent lamp (CFL). Such sources consume about one-fifth of the power for the same light output and have a lifetime almost ten times longer (according to manufacturers' specifications). Although the CFLs save energy and increase comfort, consumer resistance to the adoption of more energy efficient lighting products is high.

Many of these programs are aimed at the significant barriers that limit the efficiency of the market to capture the large potential for cost-effective electricity conservation through improved lighting technologies such as compact fluorescent lamps. One of the dominating barriers is the relatively high price of CFLs and lack of consumer awareness that the replacement of an incandescent lamp by a CFL is a profitable investment despite the relatively high price of the latter. As a result of a market study conducted by the Polish Foundation for Energy Efficiency (FEWE) [9] during the fall of 1993, a special electricity conservation program has been proposed. Its description is given below.

2. The Polish Efficient Lighting Project - PELP

In 1995, Poland received a five million USD grant from the Global Environment Facility (GEF). The project named Polish Efficient Lighting Project (PELP) was designed by the International Finance Corporation (IFC), the World Bank's branch for the private sector, the Polish Foundation for Energy Efficiency, The Institute for Energy Conservation (IIEC), and Battelle Pacific Northwest National Laboratory (PNL). The grant is administered by the International Finance Corporation, and the day-to-day running of the project is the responsibility of a Dutch company NECEL established in Warsaw.

PELP has four distinct components:

- Education 570 000 USD,
- CFL subsidy 2.7 million USD,
- Luminaries subsidy 410 000 USD, and
- Demand-Side-Management - (DSM) Pilot Project 430 000 USD.

The remaining 890 000 USD were assigned to project monitoring, assessment, and management.

In this paper we describe the DSM-Pilot part of PELP and report its preliminary results.

Before coming to this particular issue, let us first briefly describe the main features of the whole Project PELP. The main idea behind the Project was to promote CFL sales by lowering the retail-consumer price using a subsidy scheme accompanied by an extensive marketing and education campaign. The focus of the project is CFL lamps, ballasts, and luminaries manufactured in Poland. The project embraces all Polish manufacturers of these technologies that can provide a high quality product with reliable performance to meet project specifications.

The main difference between the traditional programs promoting energy efficient electrical equipment is that the subsidy is applied to the product "at the manufacturers' gate". Shops typically calculate retail prices by multiplying the wholesale price or the preceding retail price by a marked-up percentage. Therefore, by subsidizing the manufacturer rather than the consumer, the PELP subsidy has a multiplicative effect, by which the price is reduced at each stage of the sales chain from the producer to the final retailer. Consequently, a bigger reduction of consumer price is achieved. Additionally, this procedure also reduces the VAT. The results of a similar program in California showed that 1 USD of manufacturer subsidy provides about 3 USD in consumer price reduction.

All manufactures producing CFLs and/or the dedicated luminaries in Poland are eligible to take part in the Project. The producers are assumed to pursue competitively their business objectives and have the freedom to use PELP subsidies in the most efficient way to maximize their sales and attract customers.

In order to qualify, the CFL products had to meet certain minimum quality and performance standards, thus ensuring that the program avoids damaging the market with products that fail or perform poorly. Each participating manufacturer is also required to make a commitment to additional wholesale price reductions, an additional GEF-generated benefit. All participating companies have to provide a mechanism, through their product distribution and marketing network, to control retail pricing so that the full benefit of the subsidy accrues to the consumer. Newly emerging manufacturers in the Polish market could also qualify to participate in the program.

The project has provided subsidies over a two year period to Polish domestic manufacturers of CFLs and luminaries that are ballasted to accept CFL bulbs. About 1.2 million CFLs have been subsidized through the program for Polish residential customers.

A strong market education program has been central to the project. The Polish school system has been employed to educate students about efficient lighting practices and technologies.

Education and training initiatives have been undertaken by the Association of Polish Electrical Engineers and by the Warsaw Technical University to raise the capability of Polish engineers and architects to design efficient lighting systems and employ efficient lighting technologies. Finally, a program marketing effort has employed radio, TV, and print media campaigns to educate consumers about the economic and environmental advantages of CFL technology, and their appropriate use.

The program contains also a strong price and sales monitoring and evaluation component. It has been assumed that any manufacturer not able to deliver substantial retail price discounts as a result of the subsidy by controlling retail pricing, and thus ensure maximum consumer surplus, would be removed from the program.

Estimates of the local environmental impacts were produced for two scenarios [13]. Scenario A only includes the effects of the CFLs involved in the manufacturer subsidy program, about 1.1 million lamps over three years. Scenario B includes not only the lamps involved in the subsidy program, but also the additional lamp sales induced by the program, about 5.5 million lamp over twenty years.

The assumptions used to estimate the avoided electricity generation from the project were that incandescent lamps with an average power of 75 W, will be replaced by 18 W CFLs, operating 4.4 hours per day. In addition, 10% electric system distribution losses were assumed [10]. The estimates of electricity generation avoided by the project, for both Scenario A and B, are shown in Table 1, to be compared with the total consumption of 139 TWh in 1994.

TABLE 1. Estimates of electricity generation avoided by the PELP project as described in text [GWh]

Year	Scenario A	Scenario B
1995	23,6	42,0
1996	70,7	125,0
1997	94,3	207,3
1998	94,3	292,3
1999	94,3	374,4
2000	82,5	438,2
2001	47,1	456,7
2002	11,8	450,6
2003		437,6
2004		423,0
2005		400,6
2006		375,5
2007		342,0
2008		300,0
2009		248,3
2010		192,4

3. PELP-DSM Pilot Experiment

The initial aim of the DSM-Part of PELP was to demonstrate the benefits to a utility of avoiding the costs of upgrading the power supply system by using a DSM action leading to a reduction of the power peak load. In the project preparation phase, several utilities were considered as candidates. However, it turned out that the readiness of the power utilities to get involved in a program that reduces the demand for electric power or energy consumption was very limited. In fact, Poland has at present a significant over-capacity reaching ca. 30%. This is mostly the result of drastic price increases that took place in 1989 and the early nineties, and from the changes in the Polish industry structure. In this situation, the utilities consider suicidal any action aimed at further reducing the demand for power or energy. In the absence of legal regulations that would encourage, or even enable, the utilities to take the path of IRP, this has been a rational attitude. One may hope that this conservative tradition will change in the near future. Indeed, the new Energy Act containing DSM/IRP/TPA provisions has been passed by the Polish Parliament very recently. Still the understanding of its implications for the utilities is far from being high in the Polish power sector.

Therefore, it has been decided to try to find allies among the municipal governments who are concerned with electricity savings, due to the high energy bills paid by their voters and by the city administrations themselves (for schools, hospitals, and the governments' own office buildings).

To identify the most interested candidates, FEWE asked for help from the Association of Polish Municipalities, the "Polish Network Energie-Cites" (PNEC). A letter explaining the idea of the DSM-Pilot was been sent to the 30-member cities of PNEC and a number of interesting suggestions were received. Upon careful consideration, three cities were chosen for the experiment: Chelmno, Elk, and Zywiec. The selection procedure focused on:

- electrical power supply difficulties in the town or its selected areas, resulting from limited transmission capacity (long supply lines, insufficient cable sections, overloaded transformer units, etc.); and
- involvement of the municipal authorities and power utilities in the implementation of the program.

Chelmno, a city of 22000 inhabitants, is situated in central-north Poland between Gdansk and Torun. The city officials made it clear that they were enthusiastic about implementing such a program and were ready to begin soon. The other important factor was the exceptionally positive attitude and high degree of understanding of the IRP/DSM philosophy among the management of the Power Distribution Company in Torun, notably of its director for technical matters (Chelmno lies within the service area of the utility in Torun). This detail is worth emphasizing as it provided a break-through for the project start-up and shows that new ideas have begun to penetrate the "grass-roots" level of the Polish power community. The program that would target areas of the city that have distribution capacity constraints was drafted. The city described two places where existing aluminum cables did not allow the city to supply any more power and a DSM project could help them avoid or postpone the expensive investment.

The first of the two areas is situated in the old city center, which has about 4000 inhabitants and about 2500 dwellings. Total peak demand is roughly 5 MW. More power is needed to convert the heating of some of the buildings from coal to electricity to reduce the "high-emission problem" in the city center, but this was constrained by distribution capacity. The second area was a housing establishment of typical residential blocks of flats.

The local officials, however, insisted on extending the program to the entire city. The city was small enough to allow such a program within the available funds. Therefore it was decided to adopt this solution, as it would allow one to test the effect of efforts concentrated in a single city.

Elk, a city of 54 000 inhabitants is situated in northeast Poland in the Mazurian Lakes Area. The local government representatives were also enthusiastic about proceeding with a DSM program. As in Chelmno, the technical director of the Power Distribution Company in Bialystok that supplies power to Elk was ready to cooperate and understood the problem. The city described two areas. The first one was in the city center where the local utility had refused additional hookups because the system had already reached maximum capacity. The area had about 660 dwellings and was inhabited mostly by poor people. The representatives from Elk proposed that the project be designed to achieve distribution system capacity savings sufficient to allow the conversion of some heating systems in the buildings to be renovated in this area from coal to electric resistance heating (natural gas service is not available in central Elk, thus electric heating is the only cost-effective alternative to heating with coal). Doing so would allow the city to demonstrate a means of achieving a significant reduction in local air pollutant emissions through CFL installations. The second area was located at the outskirts of the city and contained 1100 single family houses.

Zywiec, 32 000 inhabitants, is situated in south-central Poland about 100 km south-west from Kraków. City representatives have identified 12 regions where voltage drops begin to occur in peak hours and where local DSM might solve the problem. The potentially troubled areas were primarily located at the end of the distribution feeders. Two such areas were finally chosen. The total number of single family houses in both groups was about 360.

The objective of this phase of the PELP project being currently carried out is to achieve an immediate and very well-documented (number, capacity, type, manufacturer) "saturation" of households and public utilities (i.e. schools) with CFL lamps. This phase of the project will result in an assessment of the market capacity for specific CFL products, in the context of well determined promotion programs and the awareness and financial possibilities of defined customer groups.

The next issue to be investigated, having a more technical aspect, is the qualitative and quantitative evaluation of the impacts of this specific, massive, wide-area and nonlinear load on the supply network, with particular attention to:

- an overall reduction of the energy supplied from the power grid;
- a reduction of the morning and evening power demand peaks;
- a reduction in current (and, in consequence, in transformer loads) and voltage drops in wide-area, low-voltage power supply networks;
- the impact of the CFL operation on the reactive power taken from the supply network;

- the influence of the CFL operation on the quality of electrical power, including parameters such as: higher harmonics, phase asymmetry, PEN (neutral wire) current, etc.; and
- measurement of selected parameters of the CFL units available throughout the retail trade: power, higher harmonics, peak start-up current, comparison of CFL light output with traditional incandescent bulbs, etc.

The project is currently in progress and focuses on the analysis and comparison of new and existing data. The structure of the project can be divided into five major components:

1) Load research program
2) Promotion and education
3) Distributed Utilities Analysis
4) DSM distribution program
5) Direct Installation Project

The **load research program** was designed by a team of experts and researches from the Battelle-Pacific Northwest National Laboratory USA, University of Mining and Metallurgy (UMM) in Kraków, and the Polish Foundation for Energy Efficiency with assistance of experts from the power utilities in Torun (Chelmno) and Bialystok (Elk). In this paper we describe in more detail the project part for Chelmno.

The **promotion and education** campaign contained five elements:

- Advertising in media. This campaign was limited to the local TV, radio, and press because acccss to national media was too expensive.
- Meetings with residents of the project areas.
- Cooperation with local school administration which was secured by the local government. In particular, simple examples of calculations were given to the students, such that they could be easily conveyed to their parents.
- The door-to-door visits that included: distribution of a questionnaire to investigate the size of the demand for CFLs, distribution of relevant information materials and finally coupons, allowing the residents to purchase the subsidized CFLs.

The **Distributed Utilities Analysis** (**DUA**) or Local Integrated Resource Planing (LIRP) is still a new notion for most of the utilities in Poland. An additional difficulty in Poland is the lack of sufficient expertise in performing such analyses, including the data collection phase.

The input data needed include the physical characteristics of the distribution systems analyzed, the cost of their upgrading, and of the technological alternatives, especially on the demand side. Part of the necessary information will be deduced from the present load research. However, a substantial amount of data will have to be obtained from other sources. The data on load flows may require consultation at a higher level of the power utility structure. A potential barrier may be the confidentiality of some pieces of information needed. Therefore excellent cooperation with power distribution companies and power utilities in Torun and Bialystok has been critical for the success of the project.

The two DU Analyses performed for Chelmno and Elk will be of a pilot character in Poland. From the very beginning FEWE and UMM have involved the local utility employees and city administration officials in this work, by introducing them to the methodology and providing the necessary theoretical training.

Regarding the **DSM Distribution Program**, the original assumption was that the lamps would be distributed on a leasing basis, using the utilities' billing systems. In particular it was intended to use the experience of other countries, notably that of Guadeloupe, where a very successful CFL program has been implemented. It was also assumed that the leasing program based on the PELP-DSM Pilot experience would be a model for other Polish cities and possibly for cities in the neighboring countries, like Ukraine or Lithuania, which have similar past experience and present needs. More details about this part of the program are given in the next section.

The **Direct installation project** in Zywiec was designed as a rather small supplement to the experiments in Chelmno and Elk. From the technical point of view it was intended to achieve a very high saturation of the target areas with CFLs in the case when people in Chelmno and Elk would respond to the Project in a lukewarm way and the required degree of saturation would not be achieved there. This was to ensure, that at least on a small scale of about 300 households, the effect of massive CFL installation will be discernible in the evening peak reduction and harmonics distortion.

In the following chapters we describe in more detail the experiment in Chelmno which began first, and for which the initial results are enough complete to draw the first conclusions.

4. Marketing Component

A wide spectrum of lamps ranging between 7-28 W, with either electronic or magnetic ballasts was offered by four CFL manufacturers:

- Philips Lighting Poland
- General Electric
- Maya
- OSRAM

(the latter was admitted only to the DSM Pilot part of PELP as an exception, as it does not manufacture the lamps in Poland). For the need of the experiment, the price of CFL lamps in the selected towns has been significantly reduced by subsidies. The sales were based on discount coupons issued only to residents (one per household) of Elk and Chelmno, and to households in the areas of the study in Zywiec. The goals of using the coupon system of CFL distribution were:

- assuring that the subsidized lamps are not massively purchased by wholesalers for reselling them with profit outside the target cities or areas;

- allowing precise determination of the number of CFL lamps and the total sum of their power in particular districts of the towns;
- providing simultaneous information concerning the project's logistics (dates of sales and locations of the distribution points); and
- marketing the product by showing the respective price reduction.

Each coupon entitled the customer-resident of the target city to purchase up to 10 subsidized CFLs. In order to facilitate the purchase and to make the consumers decisions more optimal the coupons were subdivided into three parts allowing the customer to buy up to 2, 3 and 5 CFLs, respectively.

An additional goal of the coupons issued in the areas where physical measurements were planned was to prompt the customers to purchase the lamps in a possibly short time interval (of two weeks). Therefore, a very high price reduction was offered to residents of those areas which was applicable only during the first two weeks of the promotion sales. The number of lamps offered at this very low price was five per household. In order to secure at least a minimum reasonable saturation, a special discount was offered for two lamps per household (coupon A); the additional three lamps could be purchased at a somewhat higher price (coupon B), but still very attractive. After the two weeks of the special promotion for the monitored areas their residents could still use the A and B coupons, however, the price they had to pay then was that of coupon C available to all inhabitants (households) in the city.

The structure of the discounts (related to the average retail price outside the PELP project is given in Table 2).

TABLE 2. CFL price reductions in the PELP-DSM Pilot in Chelmno and Elk

Coupon	Price reduction	Area	Validity
A	50 - 70 %	only measured area	only the first two weeks
B	35 - 60 %	only measured area	only the first two weeks
C	25 - 50 %	the whole target city*	6 weeks of promotion

* *In Zywiec coupons C were offered only in the measurement areas.*

The coupon system, supported additionally by the marketing and education campaign seems to have worked very well. Indeed, as shown in Figure1, the sales in Chelmno achieved a huge success. The figure shows the daily sales over the promotion period. A striking feature is that, on average, sales grew until the abrupt end of the promotion, i.e. until the limit of the project CFLs had been exhausted (the residual sales on the last three days were only for those customers who couldn't buy the kind of CFL they had wanted, because they had not been temporarily available in the shop). It is remarkable to notice that the **daily sales** exceeded by far the **yearly sales** in the whole Chelmno area.

Of course, one has to take into account the special character of the sales and the additional costs related to them which were covered from the PELP-DSM budget (promotion, printing the coupons and information materials, and their distribution, etc.). Still, the experience gathered during this projects has shown that implementation of similar DSM projects is possible without external support. Indeed, the interested cities (hopefully, in the future also the power utilities) and the housing cooperatives or their organizations may take upon themselves the efforts of distributing the information materials and/or coupons (if needed). The design of such a follow-up project is in progress.

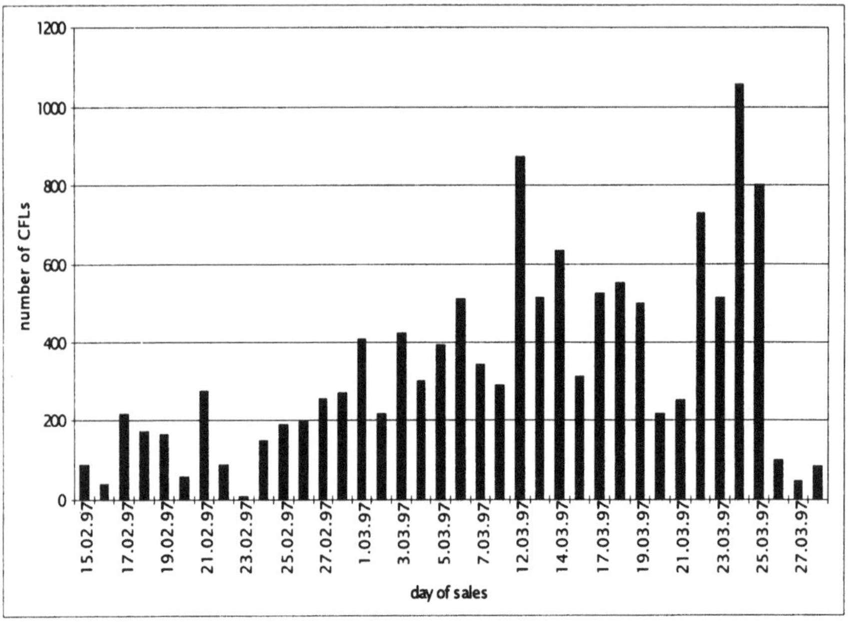

Figure 1.
Daily CFL sales in Chelmno

5. Technical Component

Special investigation zones have been set-up in each of the towns. These are areas with existing or potential power supply difficulties, and where a reduction of such problems due to massive installation of CFL units is likely to occur. Power, current, electrical power quality parameters, etc. are monitored in selected areas.

To achieve the research objectives set out in the program, a standardized scheme of the measurement system has been designed, as shown in Figure 2. The measurements are made at different points on the power distribution network (in terms of the size of the involved area and the voltage supplied to it). This allows the effect of single households on the supply network to be monitored.

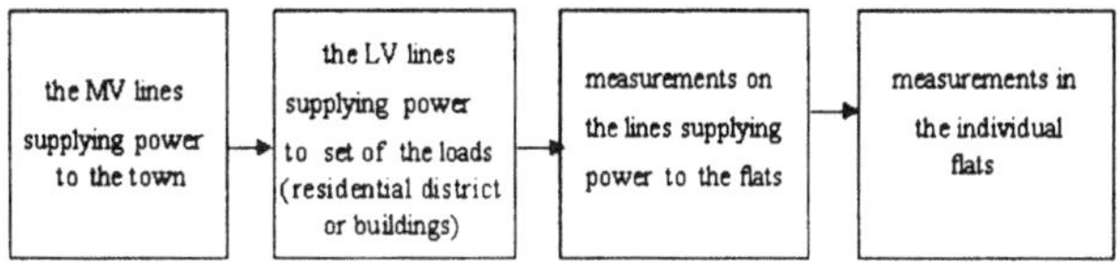

Figure 2.
Measurement system structure

The schematic diagram of the municipal power distribution network in Chelmno and the measurement points are shown in Figure 3.

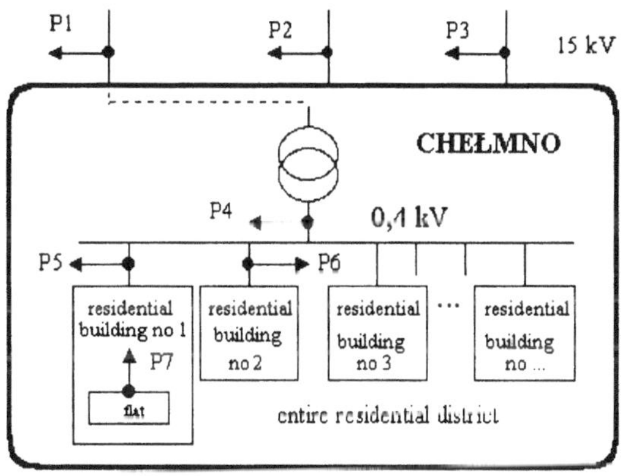

Figure 3.
Block diagram of the power distribution network in selected areas of the town

The research activities carried out within this project include the measurement and analysis of:

- transmission capacity of the supply lines;
- time of use of selected electrical devices in households; and
- electrical power quality parameters.

5.1. MEASUREMENT AND ANALYSIS OF POWER IN THE SUPPLY LINES

The equipment used in this project allows continuous monitoring of selected parameters with remote logging of 15- or 30-minutes averaged readouts. The measured parameters are the active and reactive power values monitored at the points indicated in Figure 3 (located on power lines with different voltage levels: 15 kV and 0.4 kV). At points P5 and P6 the users are residential buildings (75 flats each). Line P4 supplies an entire residential district, a small-capacity pumping station (7 kW motor), and a street lighting system (4.4 kW). The two latter uses account for a relatively insignificant fraction of the total power demand in this location. Line P1 (15kV) also supplies other residential districts and several public utility buildings (school, municipal office, etc.). Measurements of the daily variation of active power in the low-voltage-side points, indicate a significant increase in loads during the evening hours (between 4.00 p.m. and 11.00 p.m.), both during weekdays and at weekends. Figures 4 and 5 show examples of the daily load-curve shapes of active and reactive power during the selected days of the experimental period.

For simplicity, the vertical axes on the figures below are in arbitrary units, designated as %.

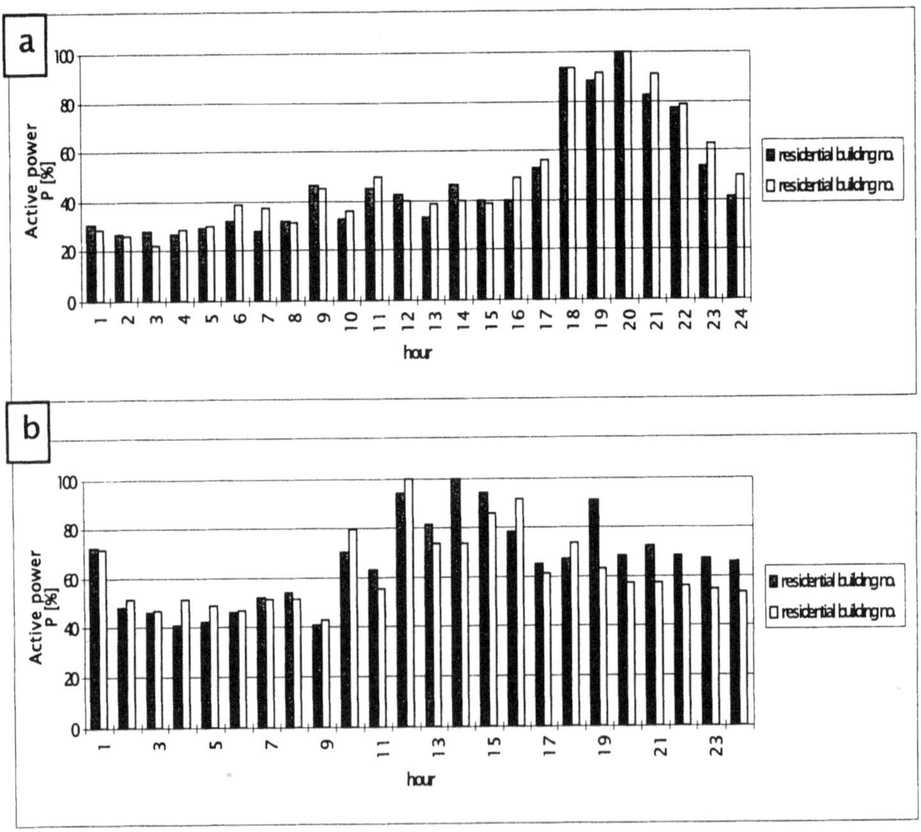

Figure 4.
Example of daily load curves of active power: (a) weekday, (b) weekend - measurement points P5 and P6 (designated as in Figure 3)

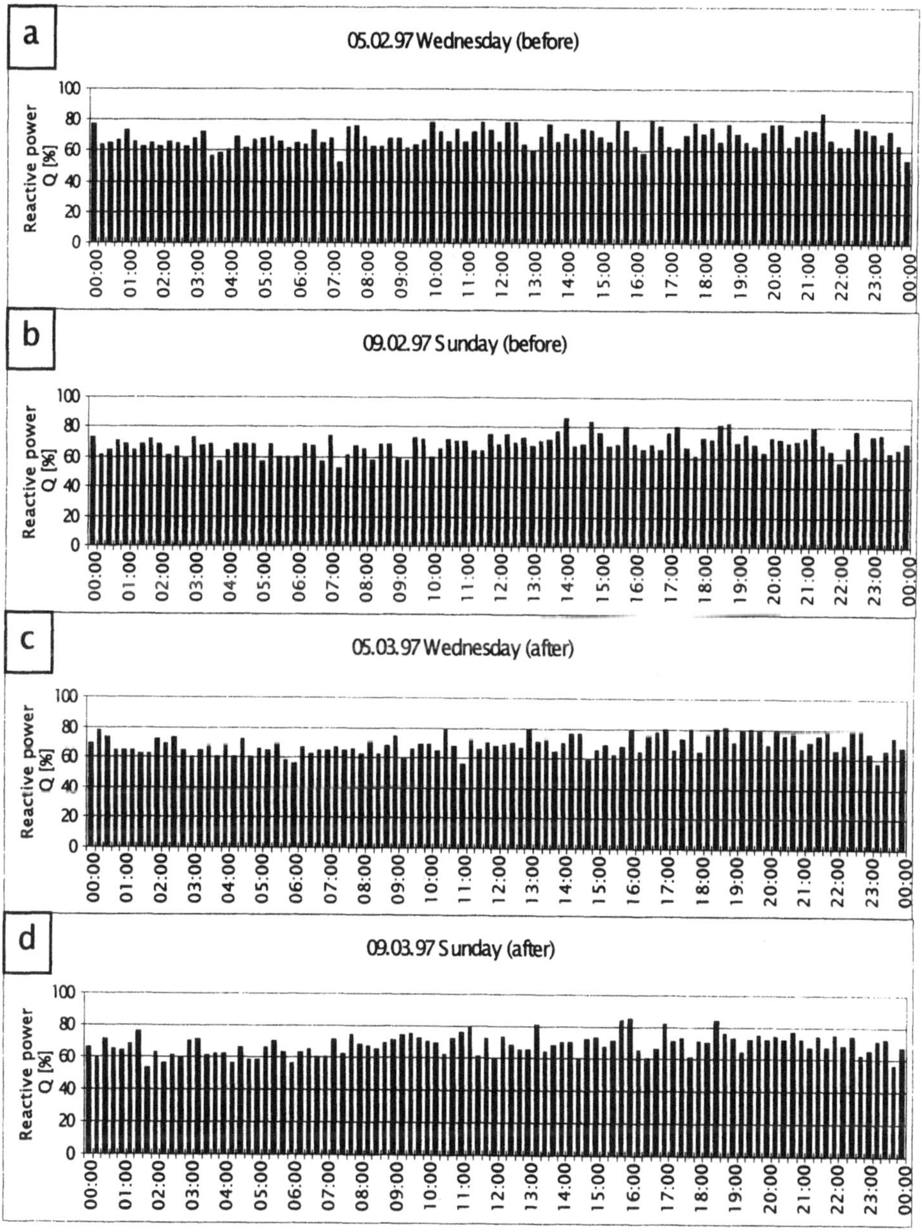

Figure 5.
Example of a 24-h reactive power-load curve shape: (a, c) - weekday; (b, d) - weekend in measurement point P5 (designations as in Figure 3) before and after the start of sales

Energy efficient CFLs have a relatively low power coefficient. Therefore the measurement system also includes registration of the reactive power to determine whether CFL installation affects the total reactive load in the supply lines.

Based on 24-h measurements of the power variations, a decision was made to perform the analysis by comparing energy consumption before and after CFL installation with the average power values during the evening peak. Calculation of this parameter takes into account the seasonal changes in daylight duration (affecting the peak schedules), and thus the home and street lighting timetables (Figure 6). Street lighting is automatically turned on by a signal from a light-level sensor.

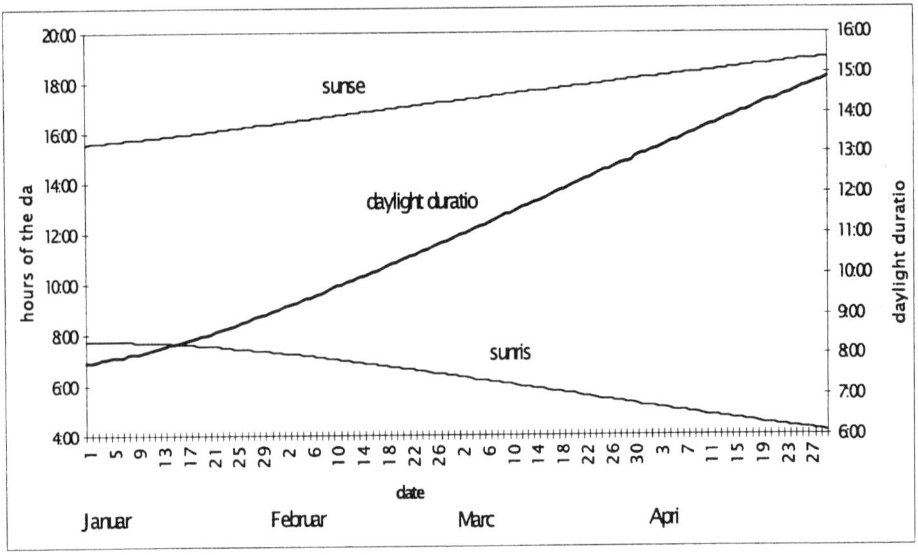

Figure 6.
Changes in sunrise and sunset hours and daylight duration within the period of time of the experiment

Figure 7 shows the changes in temperature (recorded twice a day at 5.00 p.m. and 10.00 p.m.) between January 17 and April 9, 1997. Temperature changes were not high (approximately 5°C) in the time interval of time interval of interest for the study, i.e. between 17 January and 10 March. Since space heating in the investigated buildings comes from a district plant based on coal, it did not affect the reactive power demand.

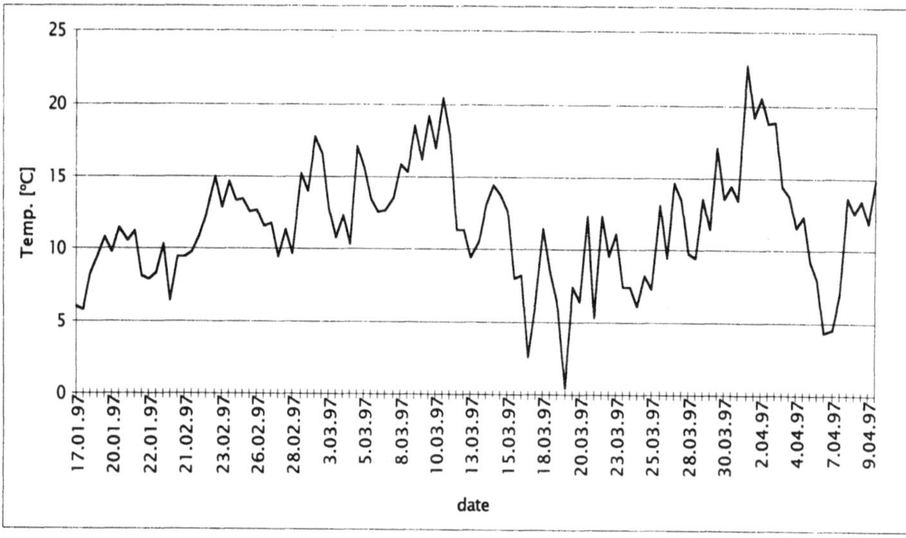

Figure 7.
Ambient temperature variations over the measurement period

Figure 8 shows the changes in average power values during the evening peak over a selected period of time. The sales of CFL started on February 15, 1997. There is a discernible reduction in power demand as a result of the installation of large numbers of compact fluorescent lamps: several thousand within the monitored area, giving an average of 5 CFLs per household.

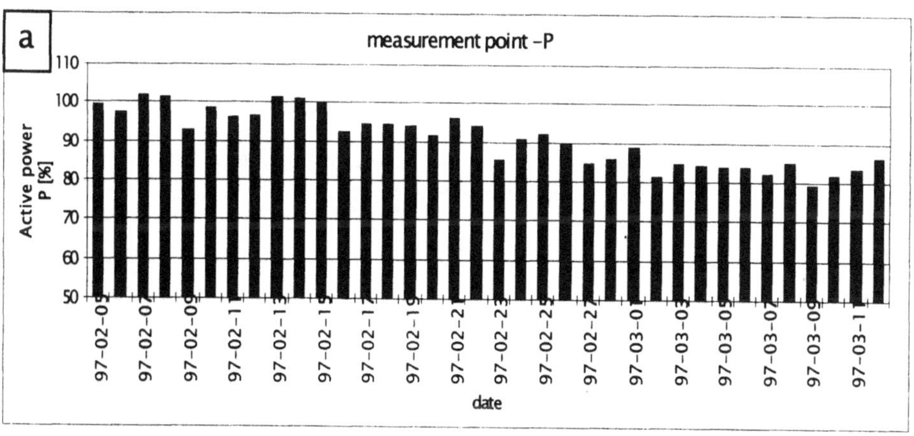

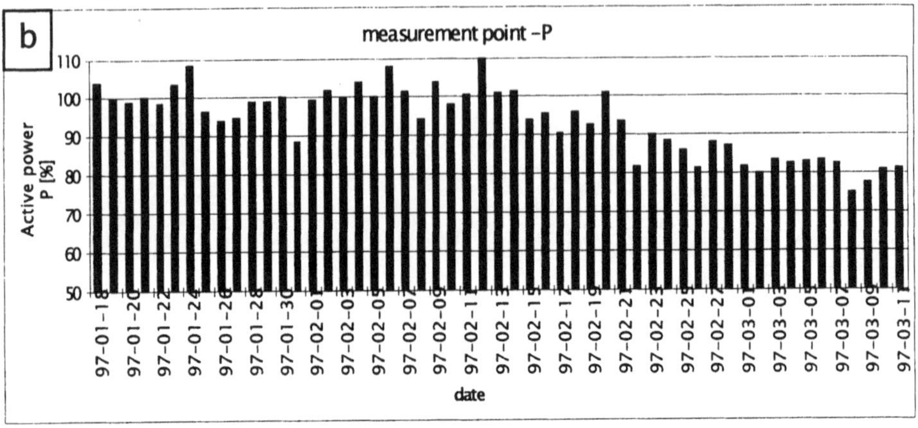

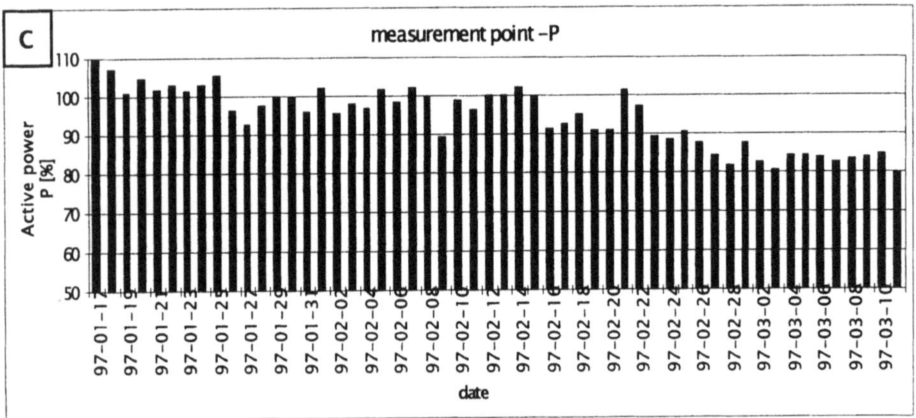

Figure 8.
Changes in average active power value (during evening peak 5.30-11.00 p. m.) at selected measurement points (designated as in Figure 3).
100 % corresponds to the average value for 17.01.-15.02.1997.

The pattern of average reactive power use during the evening peak did not change significantly between periods before and after installation of CFL lighting systems (see Figure 9). Thus, at the present stage of investigation, a hypothesis may be drawn that the effect of CFLs on total reactive power levels is negligible.

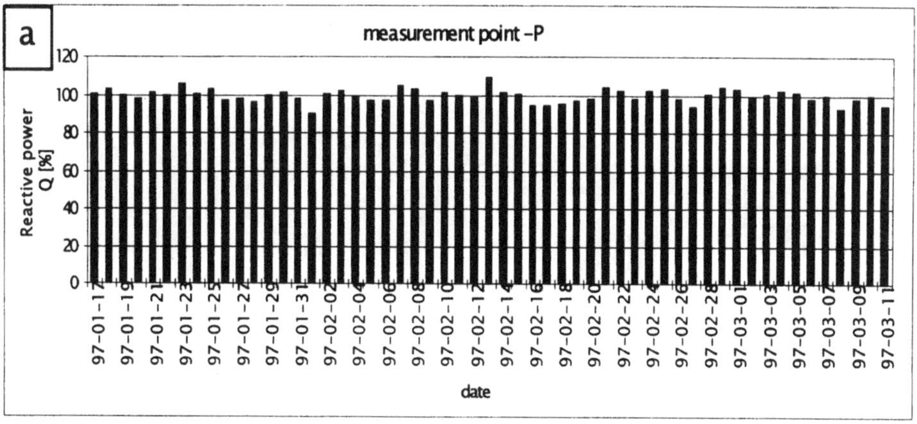

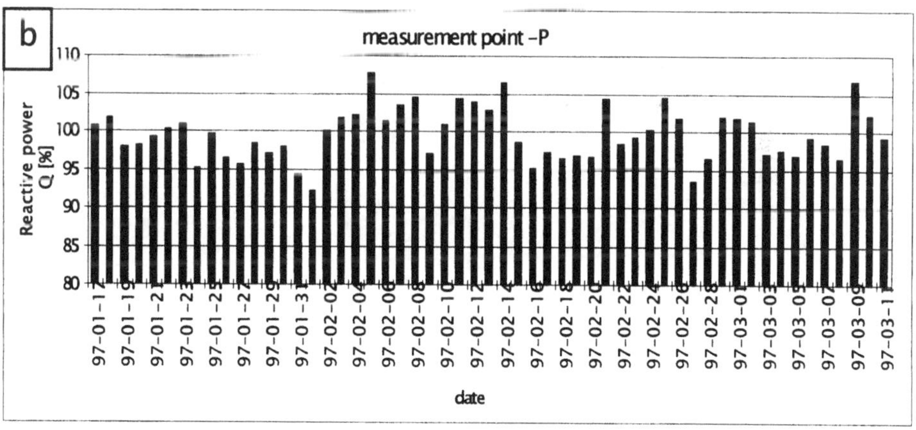

Figure 9.
Changes in average reactive power value (5.30-11.00 p. m.) in the selected measurement points (designated as in Figure 3)

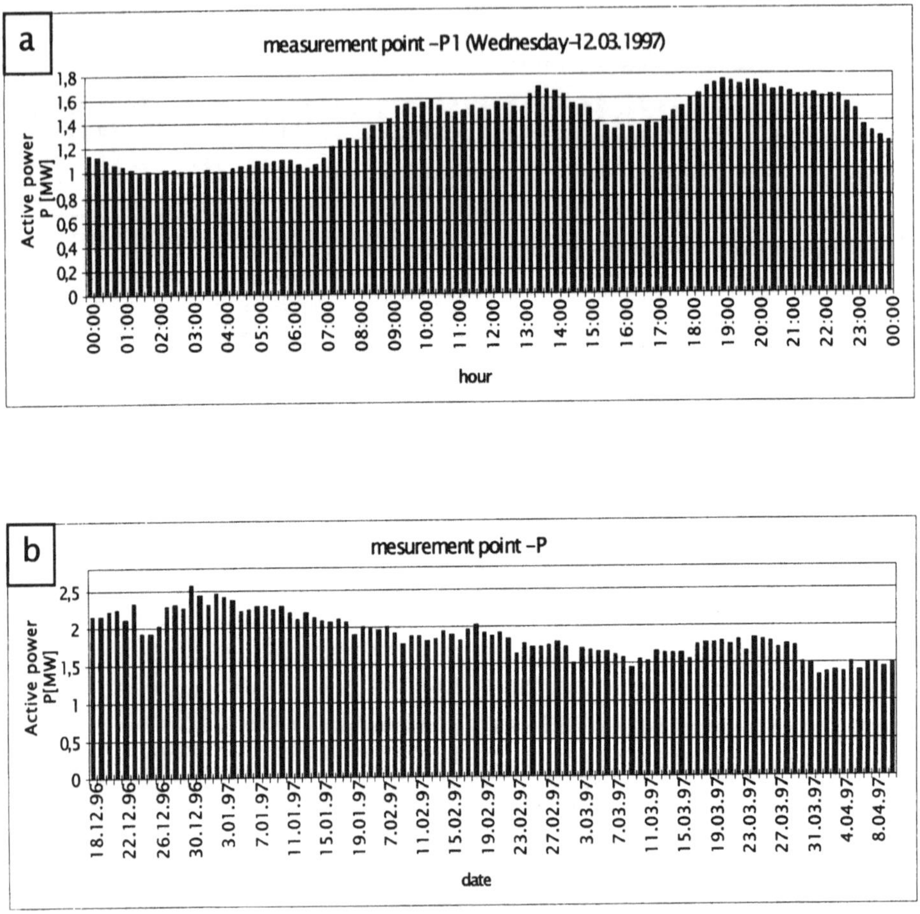

Figure 10.
Changes in active power value (5.30 - 11.00 p. m.) in point P1
(designated as in Figure 3): (a) - 24-h; (b) over a selected period of the experiment

Figure 10 shows the changes in the average power transmitted via the 15 kV line (P1). These values decreased during the course of the experiment. However, this decrease can be attributed to various uses, as the 15kV lines supply a variety of non-residential customers, and the observed effect may be result of a superposition of a multitude of factors. The complexity of different factors makes it difficult to estimate the quantitative impact of CFLs on power reduction during the evening peak at this level. Nevertheless, the number of CFL units sold and their total capacity indicate that it can have a significant impact, and an analysis of this problem has been undertaken.

At present the project is entering its concluding and data processing phase. The initial analysis of available data shows a significant reduction in active power values during the evening peaks, practically at all measurement points.

5.2. MEASUREMENT AND ANALYSIS OF THE OPERATING TIMES OF SELECTED ELECTRICAL POWER LOADS

One of the objectives of this experiment was to develop forecasting models for the electrical energy consumption in typical, socially-defined households. In Chelmno, active-power meters with automatic data logging were installed in 10 households, allowing power consumption to be monitored over the entire duration of the experiment. Additionally, data loggers recording the operating on-off times of various household appliances (refrigerators, washing machines, electrical cooking stoves, etc., and most of the lighting sources) have been installed in 90 flats (30 - for the duration of the entire experiment and 60 unit installed on a rotating basis for one week in particular flats).

Figure 11a and 11b shows an example of 24-hour change in active power in a randomly selected flat over two days: a weekday (Wednesday) and a day at the weekend (Saturday). Figure 11c shows the changes in average power (during the evening peak) over a selected period of the experiment. A clearly discernible power reduction occurs around 27 February. This is when the owner installed CFL lighting in his apartment.

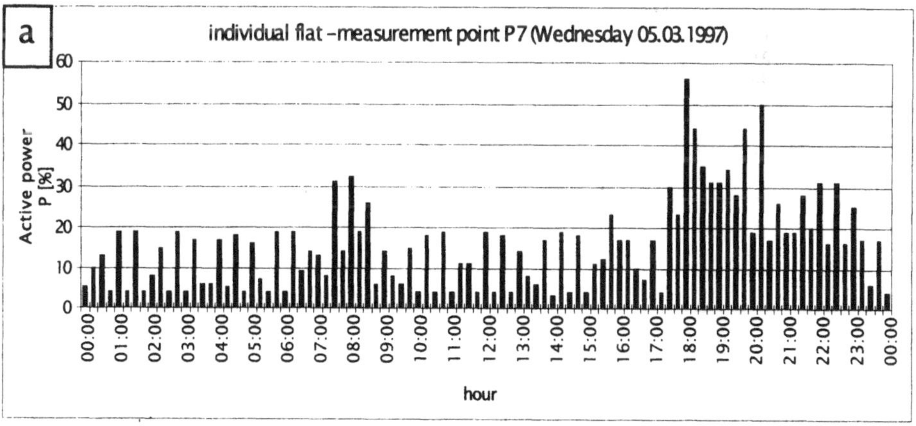

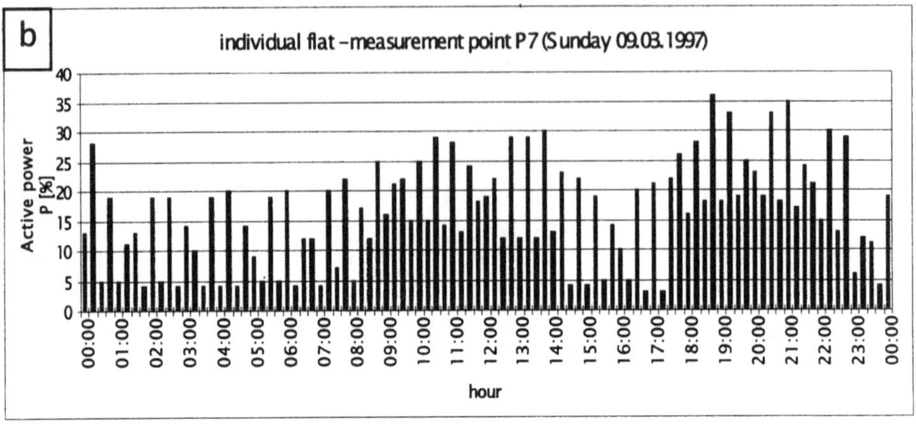

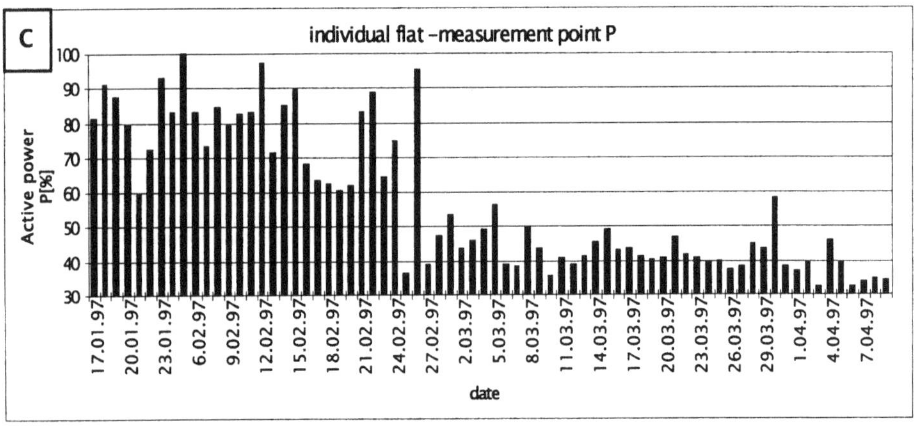

Figure 11.
Changes in average active power value (5.30 - 11.00 p. m.) in an individual flat over: (a, b) 24 hours, (c) selected period of the experiment

In this particular flat (as in many others) the operating schedules of electrical appliances were also monitored and recorded (see Figure 12). The numerical values in Figure 12 shown the operating times of selected appliances expressed as a percentage of the averaged time period: between 12.00 a.m. and 4.00 p.m., and separate readouts for all the remaining hours (individually).

Beginning of the recording: **14-02-1997** **19:15**
End of recording: **19-02-1997** **17:06**
Name of the flat owner: ....
Address: ....
Monitored receivers:

DI0:	**bathroom**
DI1:	**lighting in room No 2**
DI2:	**central lighting in room No 1**
DI3:	**side lighting in room No 1**
DI4:	**lighting in the kitchen**
DI5:	**lighting in the hall**
DI6:	**freezer**
DI7:	**refrigerator**

No,	**Date,**		**Time,**	**DI0,**	**DI1,**	**DI2,**	**DI3,**	**DI4,**
	DI5,	**DI6,**	**DI7**					
2,	**14-02-1997,**		**20:00,**	**2,**	**0,**	**20,**	**71,**	**28,**
	3,	**41,**	**17**					
3,	**14-02-1997,**		**21:00,**	**0,**	**0,**	**0,**	**100,**	**4,**
	1,	**69,**	**17**					
4,	**14-02-1997,**		**22:00,**	**1,**	**0,**	**0,**	**97,**	**23,**
	3,	**100,**	**17**					
5,	**14-02-1997,**		**23:00,**	**0,**	**0,**	**0,**	**83,**	**1,**
	0,	**12,**	**17**					
6,	**14-02-1997,**		**00:00,**	**1,**	**0,**	**0,**	**62,**	**0,**
	0,	**10,**	**17**					
7,	**15-02-1997,**		**16:00,**	**15,**	**2,**	**10,**	**12,**	**29,**
	30,	**41,**	**16,**					

Figure 12.
Data sheet from the operating time logger

5.3. MEASUREMENT AND ANALYSIS OF ENERGY QUALITY PARAMETERS

When implementing an energy-saving program other, sometimes negative, consequences of such activities should not be forgotten. This refers to many of the implemented DSM programs, and those focusing on energy-efficient CFL lighting systems. Compact fluorescent lamps are electrical loads with a strongly non-linear current-voltage characteristic. The supply current becomes strongly distorted and its waveform contains the whole range of unpaired harmonics of significant relative values (see Figure 13).

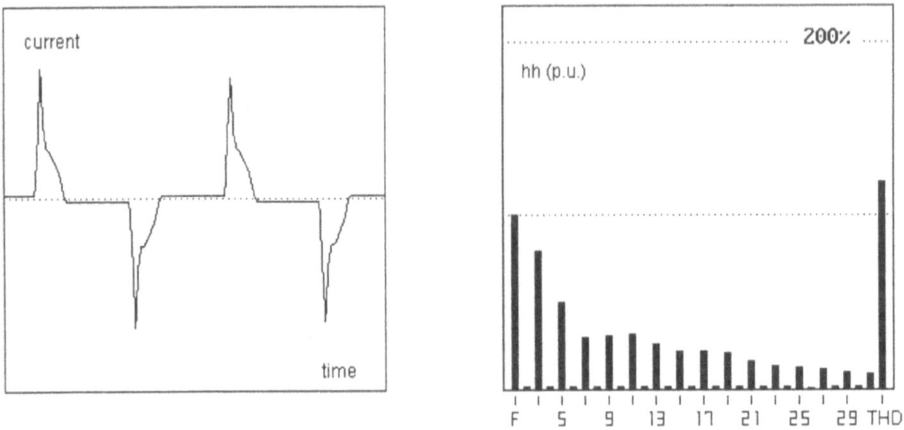

Figure 13.
Example of the current time-waveform of a compact fluorescent lamp (electronic ballast) and its harmonic spectrum.

The unit power of these sources is low, but in the case of a massive installation (in terms of numbers of units), the cumulative electromagnetic disturbances may become significant. This mainly relates to harmonic voltage distortions and the current load of the neutral wires. Presentation of the negative consequences of these phenomena falls beyond the scope of this study, but they must be recognized as significant and potentially very unfavorable from the technical and hence economic point of view.

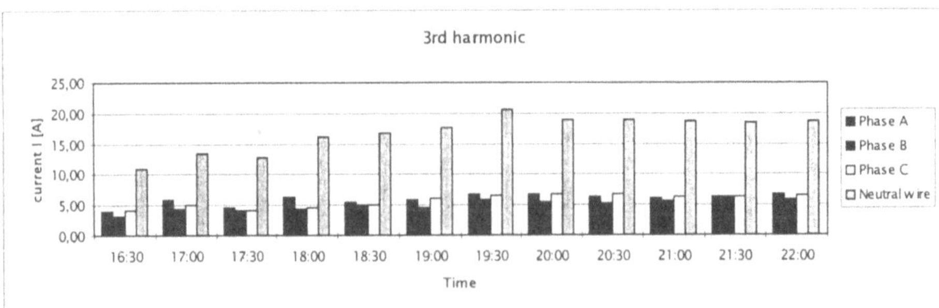

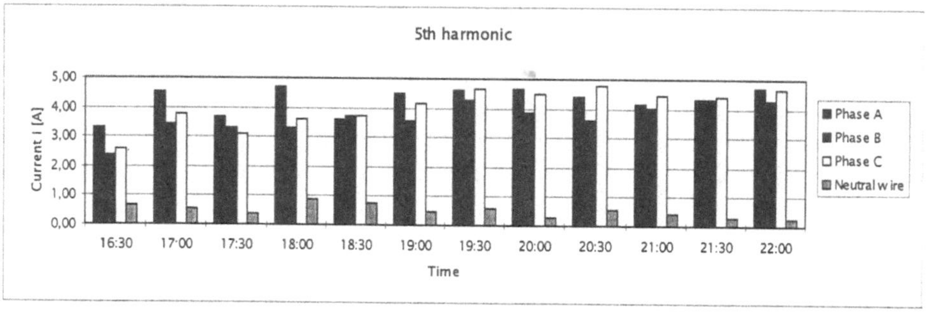

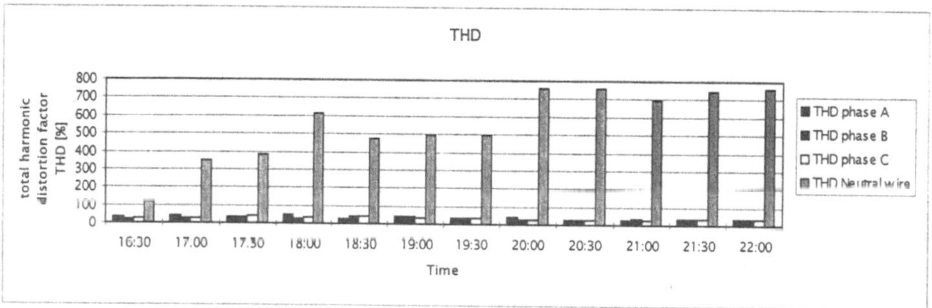

Figure 14.
Time waveform of the selected harmonics and the current total harmonic distortion factor THD in measurement point P5.

Figure 14 shows examples of the typical time waveforms of the dominating current harmonics at measurement point P5 during the evening peak. There is a discernible increase with time in the values of harmonics having orders of multiples of 3 (third is shown as an example) in the neutral wire, as people switch on consecutive electronic devices (among others the CFLs). In points P4 and P6 the waveforms are similar in character. A comparative analysis of current distortions before and after installation of CFL units is currently being performed. Its results will be very important (in terms of compatibility) of the impacts of these loads on the operation of the supply system. The preliminary results of our analysis do not show any significant increase of current higher harmonics in all phases and any significant increase of the load of the neutral wire after the massive installation of CFLs in Chelmno.

The above mentioned problem is not intended to diminish the significance of the DSM programs based on CFLs. It is only an attempt to assess their possible negative effects. It also indicates the need to consider that the outcome of DSM programs should be assessed with regard for the costs of eliminating their possible negative effects.

6. Conclusions

The DSM-Pilot experiment, constituting an important component of the Polish Efficient Lighting Project, has shown that a demand-side approach to the solution of power-supply problems is feasible in Poland and may provide very good results. The project in Chelmno seems to have achieved a remarkable success in both of its main components:

- marketing and CFL distribution,
- peak reduction (avoided power distribution costs).

The marketing aspect has been successful mainly because of the specific CFL distribution plan, based on price reduction and individual coupons, and additionally supported by a promotion and education campaign. Although such a project would not be feasible without the GEF subsidy to lamps and the financial support to the promotion/education campaign, it has a reasonable chance to be replicated or continued without an external support. Indeed, the project has demonstrated that a well designed sales (or distribution) system in which several thousands of CFLs are sold within a few weeks in one single (and relatively small) shop, can significantly lower the final consumer price. Moreover, if organised by a city or a big housing cooperative with the aim to alleviate residents' financial problems or by a local power utility to avoid or postpone power supply investment costs, the information and logistic costs can be substantially reduced. For example, a housing cooperative employees may distribute the coupons and information materials together with packages routinely delivered to their inhabitants; the meetings and education campaigns can be organized by local schools. The city or housing organization can be used as facilitators of information distribution (cable TV, information bulletins, local radio). In fact, those methods were extensively used in the present experiment.

An essential factor is also the possibility of avoiding the intermediaries by purchasing the large amounts of CFL directly at source at (hopefully) an attractive price. As mentioned above such a project is presently being designed.

On the technical side, the measurement results are encouraging. Significant reduction of the evening peak is seen at the 380 V substations in the measured target areas. Even more pronounced is the effect in the individual households. The latter, without doubt will bring palpable economic benefits to the residents (which will be publicized in the follow up campaign). The data at the 15 kV level also show a decrease. However, they still require a detailed analysis, which is in progress.

It is also important that at least no adverse effects of massive CFL installation on the power quality have been observed in this experiment at this preliminary stage of the analysis. Still, further studies are needed and will be done by a team of the University of Mining and Metallurgy and the Polish Foundation for Energy Efficiency.

7. Acknowledgments

Since cooperation of different partners was crucial in the design and implementation of the Project, the authors would like to express their gratitude to a number of people whose cooperation, contribution to this work, valuable discussions, and assistance were most essential. We thank Mr. Marc Ledbetter and Mr. Bob Pratt from Battelle PNL who guided the design of the DSM-Pilot using their huge experience and great knowledge of such programs in the United States and world-wide. Mr. Chris Granda (IFC) provided numerous remarks and guidance in shaping the marketing aspect, in particular. The cooperation and help of the power utilities managers Mr. Ryszard Michniewski (Torun) and Mr. Miroslaw Kasacki (Bialystok), cannot be underestimated. Their technical staff, in particular Mr. Wieslaw Slominski, Mr. Wieslaw Bauer, and Stanislaw Soœnicki helped to design and carry out the measurements. Also, the cooperation of the cities' governments' has to be greatly acknowledged, especially of Mr. Zdzislaw Fadrowski (mayor of Elk), Mr. Henryk Bobowski (Chelmno), and Mr. Jerzy Starypan (Zywiec). The success of the sales has to be largely attributed to Mr. Adam Puza (former mayor of Elk). We also thank our students Przemyslaw Stana, Ryszard Rudek, Pawel Rudzki, and Mariusz Filipowicz for their valuable assistance. Finally, we should acknowledge the significant financial and in-kind support of the University of Mining and Metallurgy which has made possible the extension and continuation of the research initiated in this project (UMM Grant no. 21.210.07).

8. References

1. Arseneau R., Ouellette M. (1993) The effects of supply harmonics on the performance of compact fluorescent lamps, *IEEE Trans. on Power Delivery* 8, 2.
2. Brauner G., Wimmer K. (1995) Netzruckwirkungen durch Kompaktleichtstofflampen in Niederspannungsnetzen, Technische Universitat Vien.
3. Chazottes B., Fauguembergue P., Lachaume J. (1994) Harmonic disturbances caused by compact fluorescent lamps, PQA '94 Conference Amsterdam.
4. Cidras J., Carrillo C., Arrillaga J. (1996) An iterative algorithm for the analysis of the harmonic currents produced by fluorescent lamps, 7th International Conference on Harmonics and Quality of Power, Las Vegas (USA).
5. Dwyer R., Khan A., McClusky R., Sung R. (1995) Evaluation of harmonic impacts from compact fluorescent lights on distribution systems, *IEEE Trans. on Power Systems*, 10, 4.
6. Pacific Northwest National Laboratory and the Polish Foundation for Energy Efficiency (April, 1996) DSM Program Description: Polish Efficient Lighting Project, Report, Portland (USA).

7. Emanuel A. E., Gentile T. J., Pileggi D. J., Gulachenski E. M., Root C. E. (1992) The effect of modern compact fluorescent lights on voltage distortion, IEEE/PES Summer Meeting.
8. Etezadi-Amoli M., Florence T. (1989) Power factor and harmonic distortion characteristics of energy efficient lamps, *IEEE Trans. on Power Delivery* **4**, 3.
9. Hille E., et al. (1993) FEWE Report, Warsaw.
10. IEC 1000-3-2 (1995) Electromagnetic Compatibility (EMC) - Part 3: Limits - Section 2: Limits for harmonic current emissions (equipment input current ≤ 16A per phase).
11. International Institute for Energy Conservation et. al. (1994) Developing the market for energy efficient lighting in Poland, Report for IFC, Washington, DC.
12. de Jongh K. H., Bartels J. A. (1994) Recommendation for large lighting systems to prevent harmonic distortion, PQA '94 Conference, Amsterdam.
13. Lachaume J., Fauguembergue P. (Oct. 19-22, 1993) Harmonic in distribution systems generated by flurocompact lamps, 5th Alberta Exposition and Conference on Power Quality, Calgary.
14. Ledbetter M. R. et. al. (1994) Local environmental benefits of the IFC/GEF Poland Efficient Lighting Project, Report prepared for IFC by PNL, Pichland, Washington, (USA).
15. Philips lighting, Technical information.
16. Pileggi D.J., Gulachenski E. M., Root C. E., Gentile T. J., Emanuel A. E. (1993) The effect of modern compact fluorescent lights on voltage distortion, *IEEE Trans. on Power Delivery*, 8, 3.
17. do Prado R. N., Bonaldo S. A., Moreira M. C., Vidor D. L. (1996) Electronic ballast with a high power factor for fluorescent lamps, *0-7803-3500-7/96 IEEE.*
18. Siwik A. (1996) Energy-saving discharge lamps as a sources of disturbances in the low-voltage power network, *JUEE* 2, 2, (in polish).
19. Schwabe R. J., Zelingher S., Key T., Phipps K., Melhorn C. J., Williams S. (1994) Electromagnetic susceptibility of electronic devices to high efficiency electronic lighting, PQA '94 Conference Amsterdam.
20. Topalis F. V. (1993) Efficiency of energy saving lamps and harmonic distortion in distribution systems. *IEEE Trans. on Power Delivery*, 8, 4.
21. Tribwell L. S., Lerman D. I., Baseline residential lighting energy use study.
22. Zawadil B., Khan A. K., Sung R., Houy T. (1994) Evaluation of harmonic impacts from compact fluorescent lights on distribution systems, PQA '94 Conference, Amsterdam.

EFFICIENT USE AND GENERATION OF ELECTRICITY IN HUNGARY

GYÖRGY SIGMOND
Budapest District Heating Work Co.,
Kalotaszeg u. 31
H-1116 Budapest, Hungary

1. General Overview of Electricity Consumption and Electricity Sources

The electricity consumption in Hungary is at medium level compared with other countries in Europe. In 1993, the total consumption amounted to 27,4 TWh at a population of 10,3 million. Countries with similar populations have similar or higher consumption. The total available electricity at system-level accounted for 35,4 TWh in 1993, including network losses and self-consumption of the power plants.

TABLE 1. Population and electricity consumption in Hungary and in selected countries of Europe

	Population 1993 (Mio cap.)	Electricity Consumption 1993 (TWh)	GDP related Demand (USD/kWh)	Specific Demand (kWh/cap.)
Hungary	10,3	27,4	1,83	2660
Austria	8,0	44,5	0,66	5562
Belarus	10,4	30,7	-	2952
Belgium	10,1	63,5	0,71	6287
Czech Republic	10,3	43,0	-	4175
Greece	10,4	31,1	0,98	2990
Portugal	9,9	26,0	1,10	2626
Sweden	8,8	120,9	1,23	13739
Yugoslavia	10,5	25,7	-	2448
EU	371,0	1894,8	-	5107

S. Nørgård et al. (eds.), Development with Sustainable Use of Electricity, 97-102.

The specific demand on electricity in Hungary is high compared to national economic production. The GDP-related electricity consumption is twice higher than in similar Western countries. This high consumption results from the less efficient electricity use and the worst competitive position of the national economy compared to the EU countries. Nevertheless, the structure of the consumption is quite similar to other European countries. The participation share of the industrial, residential, and other consumers each account for one-third the total.

Primary energy consumption in Hungary reached 1038 PJ in 1994; electricity production represented about 40% of it. In the same year, 50% of electricity was generated by fossil fuel combustion. (The share of nuclear energy accounts for about 45%, and the import-export has a 5% positive balance.) This fossil fuel power production plays a critical role in greenhouse gas and air pollutants emission.

TABLE 2. Emission of greenhouse gases and air pollutants in Hungary (Mt/a)

	SO_X*	NO_X*	CO*	CO_2**
Public Power Plants	0,456	0,037	0,018	19,9
Total	0,841	0,183	0,836	65,5

Emitting 65,5 Mt of CO_2, Hungary contributes to 0,3% of the energy-related CO_2 global emissions. (At the Rio Summit, Hungary confirmed the year 1987, when the economic crisis began, as the reference year for greenhouse gas emission. In 1987 the amount of the CO_2 emission in Hungary was 85,9 Mt/a.)

2. Efficient Use of Electricity in Hungary

2.1. HOUSEHOLDS

Household electricity consumption recently stabilized following constant increases in previous years. The total consumption in 1992 was 10,5 TWh, more than twice levels in 1980 (5,1 TWh). The consumption over the next two years reflects the lower financial capability for modernizing households (9,7 TWh in 1993 and 9,8 TWh in 1994). Table 3 shows the composition of household consumption for 1992.

Three factors influenced this household consumption. The decreasing standard of living did not stop fully household modernization. In other words, consumers increased the quantity and types of household equipment (e.g. freezers, dishwashers). Due to import liberalization, high technology household equipment became available on the market;

however, an uncontrolled quantity of obsolete technologies appeared in the Hungarian market concurrently. For example, some types of imported refrigerators represent 1,6 times the electricity consumption offered in Hungarian products. An official labeling system will be launched soon to promote consumer awareness. The application of energy saving technologies in the households (efficient equipment, compact florescent bulbs) will be realized slowly because of financial insolvency and the low return on investment costs at actual energy prices.

TABLE 3. Electricity consumption of households in 1992 (GWh)

	Actual Consumption	Best Available Technology	Possible Savings
Heating	1060	1060**	-
Cooking	490	490**	-
Warm water	3100	2150	950
Lighting	1100	700*	400
Washing	270	220	50
Cooling	1800	1330	470
Freezing	400	310	90
Other	2242	2242**	-
Total	10462	8502	1960

* 3 bulbs per flat ** no data available

2.2. INDUSTRY AND OTHER CONSUMERS

There is an unstable equilibrium in the industrial electricity consumption. After a crisis related decreasing (14,8 TWh in 1990, however 11,7 TWh in 1991), the consumption fluctuation was lower than 0,6% over the next 3 years (1991-1994). The closing of many energy intensive industrial consumers and simultaneous establishing of new firms with modern technologies caused this decrease.

Public services and commerce represent two-thirds of the group of other consumers. A slowly decreasing consumption (1,9 TWh) was registered in the last four years due to moderate modernization in this sector. For example, public lighting units increased 8% between 1990-1994, while consumption decreased 9%. On the other hand, in 1995, there was a secondary summer peak demand due to the increased use of climatization in the public sector. However, only incomplete statistical data are available, as the private sector has absorbed an increasing share of the economy and services (more than 50%).

3. Conditions for Reducing Electricity Consumption

Because of financial problems in the state budget, the administration did not establish funds for energy savings. There are available funds with restricted capital from various Western sources (German cool aid, Phare, etc.), which secure inexpensive credits for energy savings. The credibility of venture capitalists and institutions is poor. On the other hand, there are only some private foundations for supporting consumption savings in the residential sector, with very restricted sources. The commencement of the labeling program can reduce some types of consumption.

In Hungary, electricity distribution companies and some power plants have become privatized. The new owners have not shown any interest in the application of DSM methods until now. Presently, the state is preparing a DSM program that would require electricity companies to use IRP methods when planning new capacities.

Electricity prices have increased in the last four years, but real prices decreased 50% between 1990-1995 relative to inflation. In 1996, a planned next price increase will bring electricity prices to a cost-related level. Actual prices: 6 USc/kWh (industry); 7,5 USc/kWh (residential). Following that, prices will be regulated by price rules controlled by the state control board. The higher prices will increase interest in energy efficiency, however the low credibility and solvency will allow only slow improvements of energy efficiency in all sectors of the economy without state subsidies and programs.

Modernization and use of energy saving technologies can reduce some TWhs of consumption in the next few years. However, because of an expected slow increase in the economy, an 8-10% higher electricity consumption will be forecasted for 2000.

4. Efficient Transmission and Generation of Electricity

There exists also a high potential for saving electricity in the transmission lines. Table 4 shows the network losses for some countries.

The basic network is relatively modern in Hungary. The 400 kV and 750 kV transmission comprise more than 50% of the total basic network length. A significant cause for the high network losses is the too centralized supply system and the low participation of local power plants in electricity production. Local and industrial power plants account for only 16% of the installed capacity (1220 MW out of 7650 MW, in 1996), and annual production shares are similar. For example, local power plants generated only 14,5% of the electricity consumed in Budapest (5,8 TWh in 1992). The share of independent power plants on the system capacity are low as well (only 300MW in 1996).

TABLE 4. Network losses in selected countries

Country	Net Consumption	Network Loss	
	GWh	GWh	%
Hungary	29,7	2,9	9,2
Austria	45,9	3,0	6,1
Belgium	64,7	3,6	5,2
Bulgaria	28,9	4,9	14,6
Czech Republic	48,7	3,9	7,3
Greece	33,0	2,7	7,5
Yugoslavia	27,2	3,6	11,3

There are good options for establishing new local power plants in Hungarian cities. More than one hundred cities dispose of one or more district heating network systems (340 altogether), and they represent about 86 PJ heat demand (and 110 PJ primary energy demand). Yet 36 PJ of this demand will be served by cogeneration with a very bad heat-electricity connection factor. In 1994 only 2 GWh electricity was generated in those cogeneration plants. (In 1995 and 1996, two new cogeneration plants with 140 MW_e capacity each were commissioned.)

There are about 300 DH (district heating) systems in Hungary without cogeneration. About 700 MW of new small- and medium-sized CHP (combined heat and power) capacity can be connected to this system. Moreover, 340 MW of local CHP capacity can be built connected to existing CHP plants for DH. Thus, the share of local production can rise 25% and reduce transmission losses as well.

CHP can create significant primary energy savings through decreasing heat losses of power plants. The national electricity company plans to build 800 MW of CHP capacities over the next ten years (together with the 340 MW mentioned above). This development is part of the modernization program of 3000 MW and building of 1000 MW of new capacity. If the national electric company further replaces 700 MW of current power plants with local CHP plants, the 1500 MW of new CHP capacity could reduce the primary energy demands by 20 PJ/a. That provides, together with the primary energy saving due to the increased efficiency of the modernized power plants (55 PJ), a security for the primary energy demand of the new capacities. Further savings are possible through increasing of the present 17% share of DH. There is also a large CHP potential in the industry and institution sector.

The saving of 50-100 PJ of fuel per year through modernization and cogeneration helps to avoid the emission of 3-6 Mt CO_2; however, the surplus demand creates new emissions. Unfortunately, new private owners of electricity companies are not interested in

HP and decentralized energy generation. They have more interest in building new large capacities. This situation must be handled through legislation (IRP, DSM) or through privatization contracts.

5. Summary

The electricity consumption in Hungary is moderate but contributes highly to national primary energy demand and greenhouse gas emissions. There is around a 20% potential in savings in household consumption. The savings potential in the industry depends on the market-related technology transition. A slowly modernization is occurring in other sectors. However, new available technologies present new demands (freezing, climatization) and funds for energy savings are limited. Due to privatization, only incomplete statistical data about the consumption of the private sector are available.

Some savings of electricity network losses are attainable through decentralizing the power generation. The use of cogeneration helps to avoid a significant quantity of primary energy consumption and greenhouse gases emissions.

The market economy brings new technologies but opens the market for inefficient technologies. Real energy prices promote energy savings, but there is a lack of capital for modernization. The privatization of the electricity sector brings modern technologies, but new owners have more interest in investing in new capacities rather than energy savings. State funds, IRP and DSM programs, new statistical methods for privatization contracts, and proper legislation are needed for saving more primary energy.

ELECTRICITY SAVING POLICIES IN DENMARK

ANNETTE GYDESEN
Danish Energy Agency
Amaliegade 44
DK-1256 Copenhagen K, Denmark

1. Introduction.

The Danish Energy Agency is one of four agencies in the Ministry of Environment and Energy. The Danish Energy Agency prepares and administers legislation related to production, supply and consumption of energy in Denmark. Furthermore the Agency carry out assessments of trends in energy and elaborates energy plans. Right now and in the next years one of the most important tasks is to follow up on Denmark's new energy action plan Energy 21. Energy 21 maintains the national target of reducing carbon dioxide emissions by 20 % by the year 2005 in relation to the 1988 level.
The present paper describes the energy policy carried out in Denmark from the early 1970 till today. The main stress is laid on electricity planning and the latest energy plan.

2. Energy planning in the 1970's and 1980's.

In the early 1970's Denmark was depending of external energy supply such as coal and oil imports. For instance oil accounted for over 90 % of the energy consumption in Denmark. Therefore Denmark was seriously hurt by the oil crises in 1973/74 (lack of oil on the world market) and in 1979/80 (high oil prices).
During and between the energy crises energy consumption dropped abruptly, and then after a while increased.

S. Nørgård et al. (eds.), Development with Sustainable Use of Electricity, 103-116.

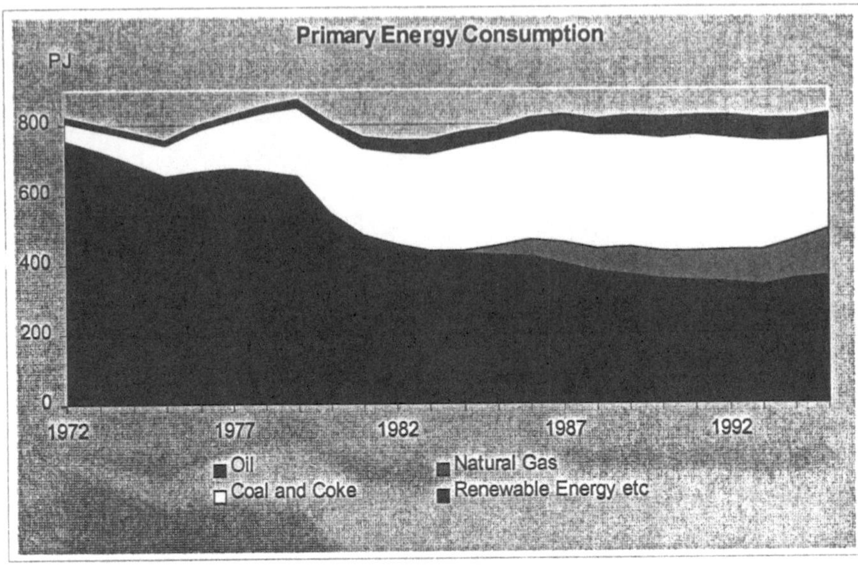

Figure 1.
Development in gross energy consumption by fuel

Before the mid 1970's the Danish energy sector was subject to only limited public regulation. The first comprehensive energy plan was Danish Energy Policy 1976. The second one was published in 1981 and nicknamed EP 81.

As a consequence of the oil crises Danish energy policy in the seventies and most of the eighties was devoted to developing our own resources, improving energy security, and increasing our self-sufficiency of energy.

A number of actions on consumption and supply were taken. Savings in heat consumption were given high priority through subsidies, information campaigns and alteration of the Building Regulations. Power plants switched from oil to coal, district heating networks were established to exploit waste heat from power plants, the use of natural gas was increased through establishment of a natural gas network covering most of the country and renewables was developed.
Taxes on oil and electricity were introduced in 1977 and on coal in 1982. When the price of oil fell drastically in 1985/86 Danish parliament decided to raise these taxes to keep energy prices outside the business sector more or less unchanged. High energy prices have been important elements in maintaining the economic base for CHP (combined heat and power) and natural gas projects and for incentives for energy conservation in the households.

2.1. RESULTS OF THE ENERGY PLANNING IN THE 1970's AND 1980's.

Since 1975 total gross energy consumption in Denmark has been reasonable constant and since the beginning of 1980 Danish self-sufficiency has been significantly increased. We are today netexporters of oil and gas and total self-sufficiency is approaching 80 %. (figure 2)

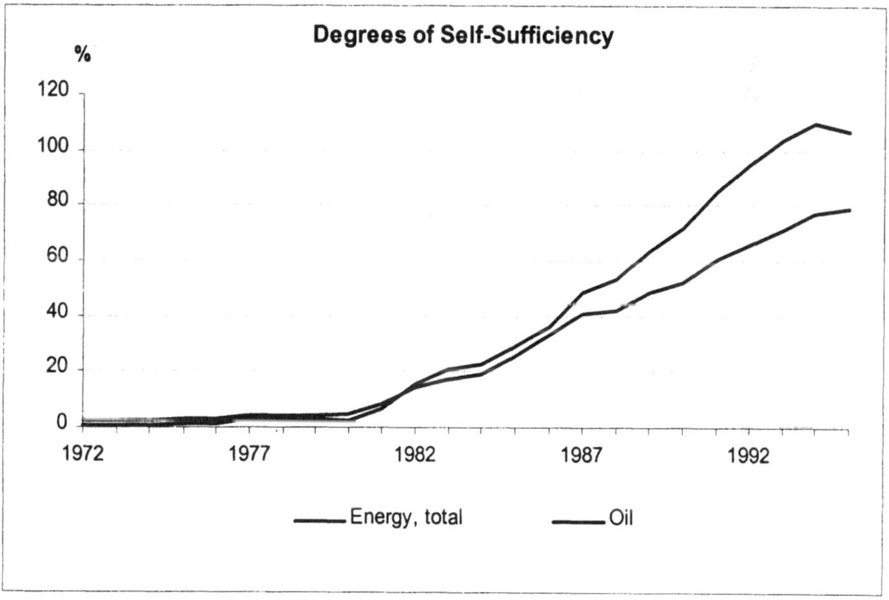

Figure 2.
Development in degrees of self-sufficiency

The gross energy consumption for space heating have been reduced by one third due to large heat savings and extension of CHP (Combined Heat and Power) networks (figure 3).

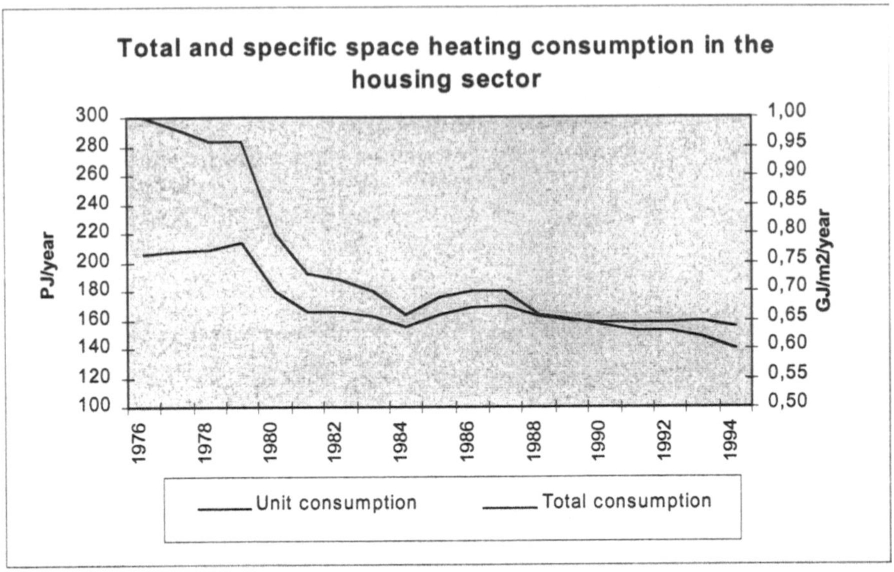

Figure 3.

On the other hand energy consumption in production (industry, agriculture, trade, services) and transport has risen.

Furthermore the electricity consumption rose steadily during the 1970's and 1980's due to an increasing coverage of various electrical appliances in the household and service sectors and an intensive automation in the production sector (figure 4).

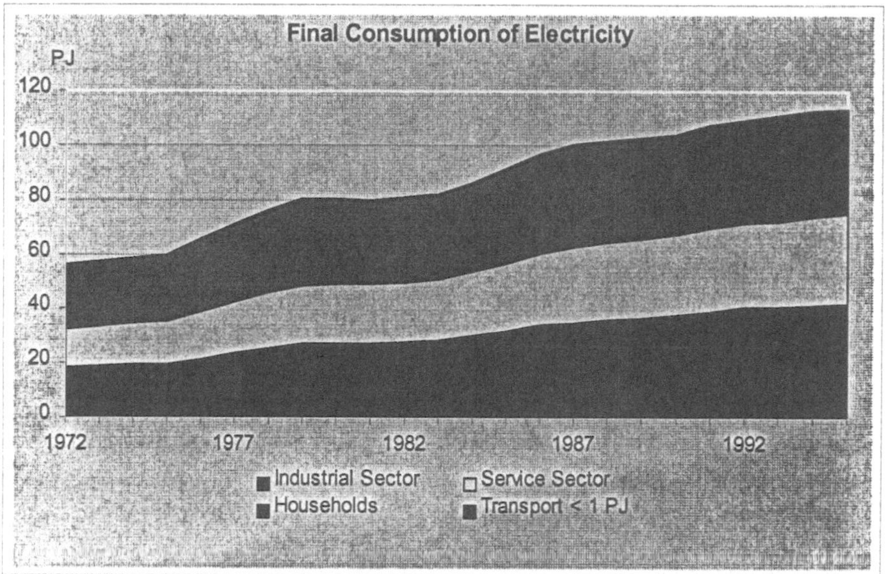

Figure 4.
Development in electricity consumption in industrial-, household-,service-, and transport sector

3. Toady's Energy Planning

Denmark is a highly industrialized country that emits large quantities of substances that burden the environment. For the energy sector, this is primarily the greenhouse gas, CO_2, where Danish per capita emissions are among the highest in the world.

Today more than 90 % of the electricity produced in Denmark takes place on coal fired power plants, which result in a very high CO_2 per produced kWh. Nearly half of the CO_2 emissions in Denmark are emitted from power plants. Therefore reduction of electricity consumption has a high priority in recent Danish energy plans.

Concern about the predicted climate changes led to submission in 1990 of the energy action plan, Energy 2000 - A Plan of Action for a Sustainable Development. Energy 2000 is still playing a central role in today's Danish energy planning.

3.1. ENERGY 2000

The main objectives in Energy 2000 was to reduce Denmark's CO_2 emission by 20 % and total energy consumption by 15 % from 1988 levels by the year 2005. These objectives were adopted by the Parliament with a large majority.

The Energy 2000 program called for action to be taken in four main areas:

- Savings in energy consumption, especially in the consumption of electricity
- Restructuring of the supply system to improve efficiency by increasing co-generation of power and heat and increasing the use of natural gas.
- Promotion of cleaner sources of energy like wind energy, biofuels, solar heat, etc.
- Research and development

Savings in electricity should be achieved by various means, most important was:

- Energy labelling scheme for common household appliances
- Minimum efficiency standards for energy consuming appliances and equipment
- Substitution of electric heating by cleaner fuels and CHP
- Green taxation (environmental taxes on CO_2 and SO_2 and introduction of environmental taxes on the commercial sector's energy consumption).

The following actions were taken :

- Denmark has been taken a very active part in international activities regarding Energy labelling and Minimum Efficiency Standards. In 1990 Denmark notified to the EU Commission an order on energy labelling which was adopted in 1992. In 1993 Denmark carried out a pilot project regarding introduction of the EU labelling scheme for refrigerators and freezers.
- With effect from 1993, a number of existing household energy taxes was converted to a CO_2 tax. that is a tax on energy consumption, levied proportional to CO_2 emissions related to the individual fuels. The taxes corresponds to about DKK 100 (about 17 US$) per tonne CO_2. Generally a reimbursement of 50 % was available to enterprises. However special reimbursements were awarded to the most energy-intensive businesses. The special reimbursements for energy intensive enterprises effectively reduce the average energy tax burden of industry to DKK 35 (about 6 US $) per tonne CO_2.
- In 1993 the parliament adopted a tax reform which among other things has the purpose to make a gradual shift of the balance of the tax system away from tax on earned income towards tax on the use of scarce resources (the so-called green taxes). The tax reform is phased in during the period 1994-1998. Overall the tax on energy for heating and power will increase from approx. DKK 450 (about 70 US $) per tonne CO_2 to approx. for DKK 600 (about 100 US$) per tonne CO_2.. For instance the green tax lead to a rise of the consumer price of electricity of about 20 %. The green taxes do not apply to energy use of enterprises. The tax reform therefore amplifies the difference between the energy tax levels of households and industry.

3.2. ENERGY 2000 - FOLLOW UP

In 1993 the government had to realize that the existing measures were not sufficient to meet the CO_2-reduction target. The gap between the predicted development and the goal was estimated to be about 5-6 %. A number of suppositions as for instance minimum efficiency standards and taxes on energy consumption in industry have either not been initiated or only partially implemented.

Especially the electricity consumption was growing faster than calculated in Energy 2000. Between 1992 and 2005, total electricity consumption was expected to rise by about 17 % if no further measures were taken.
It is in particular in manufacturing and the service sector that the increase is expected to take place following the economic growth in the years to come.

The main reason for the increasing energy consumption in the industry appeared to be the low energy prices in general for the industry, as the enterprises traditionally have been exempted from paying energy taxes. As far as the CO_2 tax mentioned earlier, the system of refunds and exemptions undermined the effectiveness of the tax.

On this basis the initiatives in the Energy 2000 follow-up, are mainly aimed at electricity consumption. The Energy 2000-follow up emphasize the importance of minimum energy efficiency standards of electricity consuming appliances and equipment, conversion from electric heating to district heating or natural gas and the use of energy and environmental taxes in the industry.

The following actions were taken:

- In 1994 Danish Parliament adopted "Act on Standards for Energy Efficiency in Energy-consuming Equipment." The law authorize the Energy Agency to lay down binding efficiency standards for electricity consuming appliances and equipment. Later in 1994 Denmark notified (to the EU Commission) minimum efficiency standards for washing machines and dishwashers.
- The green taxes on industry was adopted by the Danish Parliament in spring 1995 as a result of extensive process of analyses and negotiations with industry. The taxes are phased in gradually during the period 1996 - 2000 and the revenue is recycled to industry. In the period 1996 - 1999, DKK 1.8 billion will be earmarked for investment grants for energy saving measures in enterprises. The tax system offers certain business the option to secure lower tax rates in return for an undertaking to implement energy saving measures as part of a binding agreement with the energy authorities.

3.3. DENMARK'S ENERGY FUTURES

Denmark's Energy Futures, a discussion paper published in December 1995, contains a technical analysis of future scenarios for energy consumption and supply in Denmark.

In order to illustrate the extent of the saving potential calculations have been made according to two different sets of assumptions:

- a passive or reference scenario, and
- a maximum effort development scenario.

In the passive scenario future consumption is calculated on the assumption that there will be no development except what is inherent in the general development of technology. In the maximum effort development scenario future consumption is calculated on the basis of implementation of maximum technical savings.

Danish Energy Futures shows that there is considerable technical potential for further reduction in energy consumption both on supply and demand side. Looking at demand side, energy end-use can be reduced substantially in all sectors. The possibilities of conservation from current levels are shown in figure 5.

Figure 5. Energy efficiency potential in an active scenario 1995 - 2030

Housing, space heating	53 %
Housing, electrical appliances	56 %
Public sector	40 %
Commercial & industry, electricity	57 %
Commercial & industry, heat/process	47 %
Average	48%

In 2030 the maximum effort scenario will result in a reduction of the total energy consumption by 48 % compared to the present level.
Concerning the CO_2 emissions the scenario shows that it is possible to obtain considerable reductions. If we do nothing more than we have already done, and just replace the worn out power plant and appliances with new efficient ones, maintain the existing carbon taxes etc. the CO_2 emissions are predicted to be almost constant until about 2015. In 2030 however, CO_2 emissions will be 8 % higher than in 1994.

If we on the other hand boost the already very active energy policy we have had in Denmark, and seek to realize the maximum technological potential for improvements in energy efficiency, CO_2 emission might drop 65 % by 2030 from today's level (figure 5).

A maximum effort would require additional costs compared to the passive case due to higher investment costs.

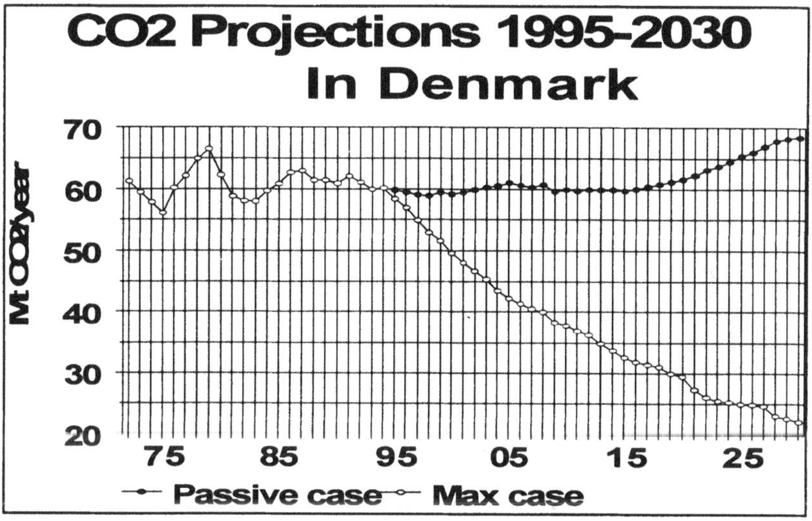

Figure 6.
The CO_2-emissions for the reference scenario (passive case) and plan scenario (max case)

We expect that the development in Denmark will lie somewhere in between the two scenario's.

Danish Energy Futures has been followed and extended by Energy 21, which lays down the energy policy agenda for the following period.

3.4. ENERGY 21

The new Danish energy action program has been published in mid April 1996. This strategy settles the strive for a long-term sustainable development of the national energy sector in a global and European context. Energy 21 includes targets for CO_2 reduction from 1988 level with

20 % in 2005
50 % in 2030

The 2030 target for a 50 % reduction is tentative and based on a scenario done by the UN climate panel IPCC in 1995. This target will be reviewed in the light of the 1996 and 1997 negotiations of the next climate protocol.

The reduction targets in Energy 21 imply that energy efficiency, CHP and utilization of renewable energy will become dominant tasks in the future.

To obtain the efficiency target, new measures are introduced in Energy 21. Realizing that the EU minimum energy efficiency standards will be delayed and do not have the expected level of ambition and that the electricity supply companies will not carry out the conversion from electric heating to CHP or natural gas additional efforts must be made.

3.5. NEW INITIATIVES IN ENERGY 21

The most important new measures introduced in Energy 21, which aims at reduce electricity consumption are:

- Establishment of an electricity saving trust
- A product oriented saving effort
- Establishment of an energy saving pool for the public sector

Electricity saving trust.
The financial resources of the electricity saving trust will be used for energy savings in households and in the public sector.

The electricity saving trust is financed through an amount of DKK 0.006 per kWh electricity consumed in household and in the public sector and payed by the consumers. A household with an average energy consumption will pay DKK 25 (about 4 US $) per year to the trust Totally this will amount to about DKK 85 mill per year or 15 mill US $.

One of the most important aims of the electricity saving trust is to support substitution of electrical heating by CHP or natural gas. However, some of the financial resources will be used for other purposes especially regarding market introduction and dissemination of efficient appliances and equipment.

Tenders for various types of electricity conservation projects will be invited from, for instance utility companies, district heating and natural gas companies, suppliers and manufacturers of electrical equipment who can bid for the projects or submit proposals of their own.

The trust is managed by a an independent committee composed of representatives of consumers, interest groups and utilities.

Act on "Electricity Saving Trust" will be presented in the parliament in the autumn 1996.

A product oriented saving effort.

In Energy 2000 - follow up minimum efficiency standards were given a central role and the Danish Government has worked hard to promote introduction of minimum efficiency standards in EU. It has, however, proved difficult to reach agreement on binding efficiency standards of the required strength. Consequently, the Government will adopt a flexible, product oriented strategy which is adapted to the current situation and the individual product.

The efforts is directed at all aspects of product's life cycle: that is research and development, production, marketing, acquisition, utilization and disposal. The instruments applied to the individual products will be adjusted regularly in step with technological development and the development of consumer demand (figure 7).

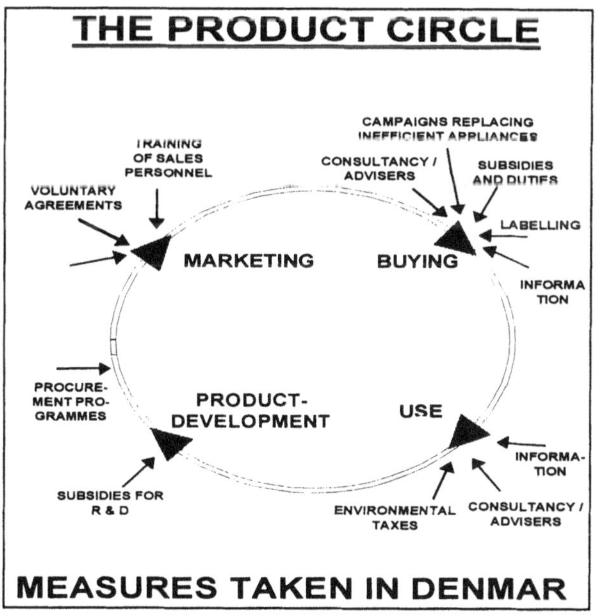

Figure 7.
Product oriented strategy

To begin with the initiatives taken will be directed towards increased efficiency of household appliances (especially refrigerators, freezers, washing machines, dishwashers, dryers, and TV/video sets), computers, office equipment, windows and heating units for housing.

The product oriented strategy includes among other things:

- energy labelling
- further initiatives for dissemination of energy efficient appliances
- procurement

Energy labelling
Energy labelling stimulates the demand for energy efficient products by making their energy consumption visible to the consumer. The implemented mandatory EU energy labelling scheme for refrigerators and freezers has demonstrated that energy labelling in combination with consumer information is a good instrument for moving sales from appliances that are relatively inefficient to more energy efficient ones. In future participation in international activities aimed at promoting dissemination of energy labelling to more groups of products will be given high priority.

Furthermore the possibilities of introducing voluntary energy labelling of products, which are not subject to other kinds of energy labelling will be explored.

Recently Denmark together with the Netherlands, Sweden, Austria and Switzerland has signed an agreement of cooperation regarding voluntary information scheme for energy efficient home electronics and office equipment to promote equipment with low stand-by power consumption.

Further initiatives for dissemination of efficient appliances
There is a large potential for cost effective conservation by ensuring that consumers buy the most energy efficient models when they replace energy consuming appliances.

The Government will therefore consider initiatives which will motivate consumers to buy the most efficient appliances. In this connection the possibilities of applying economic incentives such as subsidies that make the best appliances less costly will be analyzed.

Procurement strategies
The purpose of procurement strategies is to promote the development of and dissemination of more efficient appliances by organizing groups of buyers who can approach producers with energy efficiency demands for new products. In this way producers will become more confident that new products can be sold, and the dialogue between consumers and producer can contribute to development of products, which better reflect consumer wishes also in other areas than energy efficiency.

Initiatives will be taken to formation of groups of consumers for example in the social rented housing sector and in the business/trade and service sector.

Energy saving pool for municipalities and counties

The public sector has a special obligation to make its utilization of energy more efficient. The Government will therefore take the initiative to initiates talk with counties and municipalities with a view to establish a compulsory energy conservation pool, which will be used in full to implement energy saving measures in municipalities and counties, and will be financed by a contribution based on the energy consumption in municipalities and counties.

Further initiatives

The effort for implementation of energy efficiency standards on an international level will be continued.

3.6. EFFECTS OF ENERGY 21

Impacts of the energy plan are evaluated in relation to a reference scenario, which describes the development that is expected should the existing energy policy with associated initiatives be continued without further changes.

In both the reference and plan scenarios, the development in energy questions is based on forecasts of economic growth by about 1.7 %.

In the reference scenario the gross energy consumption rises from 1994 to 2030 by 10 %.

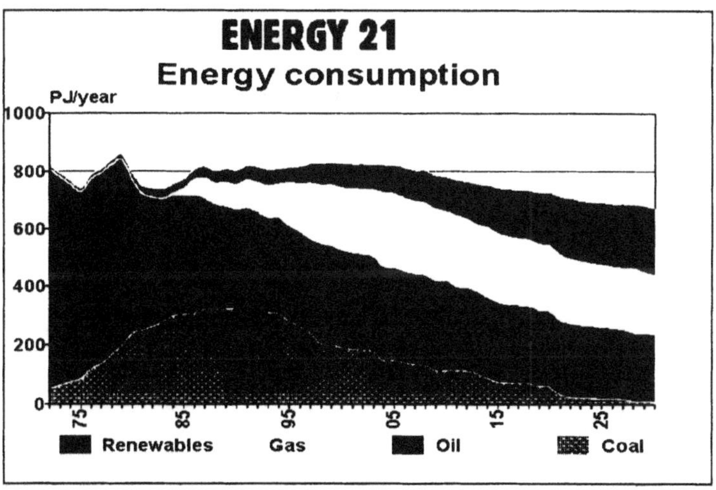

Figure 8.
Gross energy consumption in the Energy Plan scenario (Energy 21), by type of fuel.

According to the plan scenario the gross energy consumption in the period 1994-2030 falls by nearly 20 % (figure 8).

By 2005 the reference scenario leads to a reduction of CO_2 emissions of about 12 % only, while the energy plan scenario is expected to make a further reduction of about 8 %, thus achieving the required reduction of about 20 % by 2005.

4. Concluding Remarks

Compared with other countries the Danish CO_2 target is very ambitious. Only Germany has adopted a more ambitious target.

The comprehensive effort in energy policy has created substantial and visible results during the last 30 years. To mention a few:

- In existing gross energy consumption for heating has per square meter been halved.
- One third of Danish power production and almost half of the space heating are co-generated. This means more than 60 % fuel efficiency in average for the power system as a whole.
- Renewable energy has still growing importance. Wind power capacity today totals more than 10 % of the national peak load and 3 % of the power production.
- In the last years electricity consumption in the household sector has been nearly constant and in the last two years it has in fact decreased slightly. In 1994 the electricity consumption in households decreased by 1.5 %.
- From 1988 to 1994 the CO_2 emissions in Denmark have dropped about 4 % while the GDP (Gross Domestic Product) has increased by 10.5 % in real terms. In other words the development in Denmark has shown that it is possible to decouple the relation between the economic growth and the rise in energy consumption and CO_2 emissions, respectively.

LIGHTING EFFICIENCY IN RUSSIA AND CIS

JULIAN B. AIZENBERG
VNISI, Koroleva str. 3a-84
129515 Moscow
Russia

Summary

The objective of saving energy in lighting installations is of great importance in Russia and other countries of the former USSR, because in 1990, more than 13% of the total generated electric power, i.e. more than 220 TWh per year, was spent on illuminance.

The report characterizes up-to-date lighting installations and gives an analysis of the primary methods for saving electric power. Gains caused by improvements in lighting methods and installations maintenance are discussed, and associated quantitative assessments are presented.

The total electric power consumed by lighting installations in Russia can be lowered by 50% compared with today's level.

All data, which are given below for USSR (CIS), must be multiplied by 0,65 to apply to Russia only.

1. The General Problem

In the former USSR, more than 1,48 billion points of light (with one or more lamps), with a total power of over 152 GW of electrical energy were installed in 1990 in industrial, public, agricultural and domestic buildings, and city streets (Table 1). Lighting installations consumed around 220 TWh, about 13% of the electricity generated in the country. Interestingly, 54% of all generated electricity was used in industry and construction, 24% for apartments, and only 1% for exterior lighting. The lighting installations consumed practically the entire electricity generated by the nuclear power plants in the country [1].

S. Nørgård et al. (eds.), Development with Sustainable Use of Electricity, 117-125.

TABLE I.

Fields of application	Number of light points		Wattage, GW
	millions	percent	
Plants and mines lighting	800	54	90
Public buildings lighting	180	12	25
Exterior (and street) lighting	17	1	3
Domestic lighting	350	24	15
Argicultural lighting and irradiation	135	9	19

Considering the total oil, coal, and gas consumption for generation of electricity in this period, it becomes apparent that not less than 75 million tons of oil, 93 million tons of coal, and 109 billion cubic meters of gas were burned only for the purpose of lighting. As a result, 570 million tons of carbon dioxide and 80 million tons of toxic gases and ashes pollute the air every year, changing the environment in most large cities and settlements from bad to worse [2].

For lighting, about 2,5 billion of various light sources, more than 85 million luminaires, and 48 million ballast's were manufactured in the CIS.

Due to the relatively small production of high-efficiency discharge lamps and accompanying luminaires (about 150 million fluorescent lamps (FL) and 13 million high intensity discharges lamps per year, in comparison with 350 million FL and 19 million high intensity discharges lamps per year in the USA), inefficient incandescent lamps (IL) are being used in the majority of luminaires in the CIS (75% of total). As a result, the share of luminous flux of discharge lamps in lighting installations does not exceed 65%. CIS states spend about 36 kWh electricity to generate 1 million lumen hours (Mlmh) of light, while Great Britain and Japan spend 25-26 kWh electricity, i.e. 1,35 times less.

Because of the failure to systematically maintain luminaires, a minimum of fifty percent of luminaires efficiency is lost: almost 60 TWh per year, which is equivalent to 18 million tons of coal or 12 million tons of oil.

Today the situation is very different from 1990. Production of lamps and luminaires has decreased considerably (40-45% from 1990 levels). Electric energy consumption has also decreased (65-70% from 1990 levels). Hence energy stations have extra capacity and do not have interest in saving energy.

This new structure has changed the correlation between electricity consumption of different consumer categories. Today the electricity consumption in industry is drastically reduced to 28 % of total demand instead of 54 % in 1990, while it has increased in public buildings from 12 % to 25 %.

2. The Basic Scientific and Technical Directions for Increasing the Efficiency of electricity Usage

2.1. TASKS

Development of means for lighting and irradiation, and primarily, increasing the efficiency of transformation of electricity into light energy in light sources demands:

- stabilization of the light energy generated by the lamps during their lifetime (slow the decrease of luminous flux);
- manufacturing and application of luminaires with effective luminous intensity distribution curves; see figure 1a;
- decrease of the negative influence of overheating on lamp efficiency, particularly FLs and HPSLs (high pressure sodium) in luminaires;
- removing the thermal energy radiated by lamps and consequently decreasing electricity consumption for air conditioning and ventilation systems;
- decrease of energy consumption in luminaires by means of electronic ballast's;
- stabilization of luminaires characteristics in the process of maintenance through efficiency and luminous intensity distribution curves, particularly in harch environmental conditions; and
- creation and use of lighting installations with reserved capacities and dynamically controlled characteristics.

2.2. LIGHT SOURCES

It is well known that discharge lamps produce 50-100 times more light energy during their lifetime than ILs. Wide application of luminaires with more efficient light sources may save 20% to 70% on energy spent for lighting, equivalent to 300-2400 kWh electricity per one luminaires annually. The use of compact fluorescent lamps (CFL) with electronic ballast's provides 80% electricity savings. Installing only one luminaires with a CFL in 20% of apartments guarantees 3-5% energy saving in the CIS, i.e. 0,9-1,5 billion kWh per year. The widespread use of CFLs in at least 30% of today's luminaires with ILs will save 50 billion kWh electricity per year and perhaps would allow the closing of every fourth hazardous nuclear power plant.

The production and application of CFLs, however, are in poor condition. The CIS, with a population of 260 million people, manufactures 0,5 million CFLs. Many foreign companies (Philips, GE Lighting, Osram, and others) sell CFLs in Russia, but with great difficulty. The price of a CFL with a built-in electronics ballast is too high for Russian people (20-25 USD, which is 25-30% of an average monthly salary), especially when the price of electricity remains low. Table 2 shows the payback time of a 20W CFL.

The most important task in the field of light sources is to search for new ways of generating, redistributing, and applying optical radiation, which may result in qualitative changes in the use of resources, mainly in the amount of electricity used, per unit of produced light energy.

2.3. LUMINAIRES

The correct choice of luminaires ensures a large decrease in electricity consumption of a lighting installation. The luminous intensity distribution curve (LIDC) of luminaires is critical for this decrease, through its design that considerably influences the stability of operational features, through energy consumption in ballast's, as well as through the possibility of lamp reserving (figure 2). While selecting luminaires by LIDC, it is necessary to consider that luminaires with different LIDCs are needed to guarantee an efficient lighting installation. Luminaires of at least three LIDS types, for example N2, Cl and Wl [3,4], have been proven experimentally to create effective lighting installation in premises with different heights and various L/H ratios (Figure 1 and Table 3). Luminaires with energy saving FLs and effective LIDCs are being used in public buildings to remove reflected glare, improve lighting quality, and decrease lighting power from 5,5 to 2,4 W per 100 lx.

The use of luminaires with mirror reflectors, effective LIDCs, and efficient FLs in public buildings with illuminance of 1000 lx decreases the installation power necessary from 60 W/m^2 to 12.5 W/m^2.

TABLE 2. The pay-back period in years of a CFL (20 W)

Annual work, h	Price of CFL 10^{3}rub	Price of CFL USD	Price of 1 kWh, rub/cent 60/1,2	100/2	200/4	300/6	500/10S
1000	25	5	5,2	3,125	1,56	1,04	0,625
	50	10	10,4	6,25	3,12	2,08	1,25
	75	15	15,6	9,4	4,7	3,12	1,9
	100	20				4,16	2,5
	150	30					3,75
1500	25	5	3,47	2,08	1,04	0,7	0,54
	50	10	6,94	4,16	2,08	1,4	1,08
	75	15	6,24		6,24	2,1	1,62
	100	20				2,8	2,16
	150	30				4,2	3,24
2000	25	5	2,6	1,56	0,78	0,52	0,312
	50	10	5,2	3,12	1,56	1,04	0,62
	75	15		3,2	2,35	1,56	0,93
	100	20			3,12	2,08	1,25
	150	30				3,12	1,87

Considerable energy savings are possible due to the decrease in maintenance factor of the lighting installation. One of the interesting approaches in this field is the application of new devices with slit light guides (SLG). Cylindrical (or of some other shape), extended, hollow SLGs are hermetically sealed and possess aerodynamic properties that prevent pollution from affecting the optical slit. In locations with a hazardous explosive atmosphere or difficult environmental conditions, SLGs help achieve 10-15 % in energy savings.

The use of SLGs with multilamp input devices provides an additional decrease in electricity consumption. Not all the lamps should be switched on simultaneously. The number of lamps turned on may be increased gradually as the efficiency of lamps and lighting installation decreases. Electricity consumption can be reduced by 25-30 % and expenses for replacing lamps decrease dramatically as well (25-30% in energy savings).

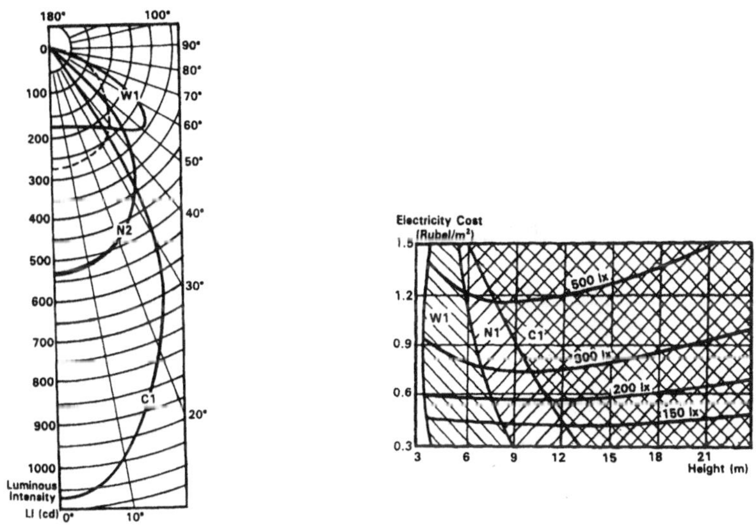

A B

Figure 1.
A shows effective luminous intensity distribution curves, LIDC, for various luminaires for industrial lighting, C1, N2 and W1. **B** shows the electricity cost for application of the luminaires with LIDC installed in different heights. The members 500 lx, etc., refer to illuminance at floor level.

TABLE 3. Energy saving provided by luminaires with effective lighting installationDC

Heights of industrial premises	Energy saving in case of replacement of luminaires with cosine LIDS with luminaires with effective LIDC	Energy saving &
5,0	N2	14
	N3	15
	C1	18
10,0	N2	22
	N3	25
	C1	28
	C2	31
15,0	N2	26
	N3	28
	C1	32
	C2	34
20,0	N2	28
	N3	32
	C1	34
	C2	38
	C3	40

Electricity savings can be achieved due to a considerable decrease in the initially designed electric power, i.e. switching on only a few of the lamps from the beginning while reserved lamps are turned on after definite time interval signifying a permissible decrease in illuminance (Figure 2).

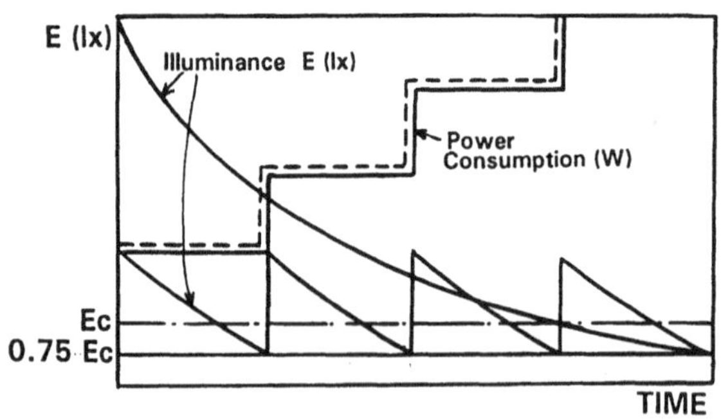

Figure 2.
Decrease in illuminance over time can be counteracted by switching on more lamps instead of starting with them all on. E_c is the illuminance but 0.75 E_c is acceptable. The area above the dotted line is saved electricity.

While designing illumination installations, for energy savings, three factors are important: Choice of lighting system; ability to autocontrol lighting; and maximal use of daylight.

Localized general lighting of industrial interiors with few irregularly mounted machines provides considerable energy savings (Table 4).

TABLE 4. Energy saving provided by localized general lighting

Mine area % of total area	Energy saving %
25	20-25
50	35-40
75	55-65

The use of general lighting combined with enhanced local lighting provides a 25-50% electricity savings.

Guidance and uniform distribution of solar radiation to the interior of buildings without daylighting is very important. Several lighting systems based on cylindrical and wedge-shape light guides are designed in Russia. In such a system, the light to be transmitted by light guides comes from daylight concentrators (heliostats) mounted on the roof or outside the building. The use of such lighting systems in the middle latitude regions may save 20-45% of electricity, i.e. 8-8,5 billion kWh per $10^8 m^2$ of conventional illuminated industrial areas (Table 5).

TABLE 5. Electricity saving from utilizing daylight and control of artificial lighting in industrial buildings in accordance with daylighting conditions.

Number of shifts	Daylighting type	Artificial lighting regulation mode	Electric energy saving
	upper lighting	permanent	36-27
1		step mode	32-13
	side lighting	permanent	22-7
		step mode	12-2
	upper lighting	permanent	31-23
2		step mode	27-11
	side lighting	permanent	19-6
		step mode	10-2

Modern sensor systems detecting infrared radiation or sound signals may turn off electrical lighting when people leave the room. Such systems provide 15-40% electricity savings.

2.4. IMPROVEMENT OF SERVICE OF LIGHTING INSTALLATIONS

There is a serious contradiction worldwide between the high technical level of modern lighting devices and the poor service of these complicated and expensive products. Permanent professional control of lighting conditions is almost absent. Developing lighting devices that do not demand technical service to design, manufacture, and sell such devices is very important. The design of lasting electrodeless lamps QL by Philips (Lifetimes = 60,000 h), while a considerable achievement, still cannot solve the maintenance problem since luminaires using these lamps become very quickly dusty and demand regular cleaning. Currently, luminaires of a new type are being worked on in Russia, which do not get dusty after years of operation. Use of such luminaires together with new lamp types could lead to impressive results.

2.5. DECREASE OF ENERGY CONSUMPTION FOR AIR CONDITIONING

An extra important source of energy saving from lighting is a decrease in the lights demand on air conditioning. In public buildings for instance, heat radiated by lighting installations creates 70% of the total radiation of heat. Decreasing the lighting installation wattage and operation time (which is achieved by automatic regulation) could allow dramatic reductions in the wattage of air conditioners. One of the best ways to decrease heat radiation, and even to utilize the heat produced by lamps, is to use lighting devices based on hollow light guides with high wattage input devices mounted on the outside the rooms to be illuminated.

2.6. GENERAL ESTIMATION OF ENERGY SAVING PROSPECT

Through 2010, the highest possible electricity savings for lighting in industrial buildings could reach 50%, considering the potential for lighting engineering progress and the real rates of lighting industry development in Russia. Interestingly, our estimation coincides with that of U.S. experts, who also believe a 50% reduction in energy consumption for lighting is possible (which means that only 10% of total energy production would be needed for lighting, instead of today's 20%).

3. Proposals of Projects That Could Be Realized in Russia

- Setting up a system of encouragement for managers who reconstruct lighting installations that provide a perceptible reduction in energy consumption.
- Setting up a favorable credit system for reconstruction of lighting systems in industrial enterprises. Credits could be paid out of the money from the saved energy.
- State energy companies could compensate partially the expenses of lamp manufacturers for shifting to CFL manufacturing, which may reduce CFL price and stimulate production and sales.
- An international body could provide credits for setting up the manufacturing of CFLs in Russia (at the desired production of 25-30 million per year), which could enable closing down the most hazardous Russian nuclear power plants without any power shortages.

4. References

1. Aizenberg Ju.B., Demirchyan K.S. and Faermark M.A. (1989) Improvements in the utilization efficiency of electric power for lighting systems, *Svetotekhnika* **12,** 1-6.
2. Makarov A.A. and Volberg D.B. (1991) *The Future of Sovjet Energetics. Evaluations on the Eve of the Transition to Market Economy,* Moscow, Institute for Energetics Research, 130.
3. Aizenberg Ju.B., Guseva L.S. and Faermark M.A., (1985) Rational utilization of luminaires with typical luminous intensity distribution curves, *Svetotekhnika* **7**, 6-8.
4. Aizenberg Ju.B. and Bukhman G.B., (1993) Method of classification for luminous intensity distribution curves and proposals of their new classification, *Lux Europa, Edinblirg* **7,** 702-707.
5. Aizenberg Ju.B., Bukhman G.B., Pjatigorskij W.M. and Korobko A.A. (1992) Beleuchtungsanlagen mit schlitzlichtleitern, *Licht* **7/8,** 624-631.
6. Lawson J. and Kwartin B. (1991) Green lights on energy saving, *LD&A*, February.

ENERGY-EFFICIENT INFORMATION AND COMMUNICATION TECHNOLOGIES IN EUROPE [1]

JACQUES ROTURIER
C.E.N.B.G.
Université Bordeaux 1
BP 120-Le Haut Vigneau
F-33175 Gradignan, France

ALAIN ANGLADE
Ademe
500 Route de Lueioles
F-06560 Valbonne
France

Abstract

In the Office Equipment electricity end-use sector, the benefits of a perfect power management were demonstrated about 10 years ago. Nowadays, the Energy Star US-EPA label announced in June 1992 is, de facto, the international standard. As a consequence, recently purchased office equipment, taking into account its typical 50 % conservation potential is expected to be twice more energy-efficient than the machines supplied in the late 80s or early 90s. However, a major obstacle, the large gap between market availability and users's standard practices, has not yet allowed to achieve that result. A better understanding of the obstacles involved, based on a monitoring campaign, funded by EC-DGXVII (PACE programme), was started in 1995 in 3 EU countries (Denmark, France and Portugal). Among many aims, the main purpose of the campaign was to obtain a realistic view of the present situation in various office activities, assessing both impediments and possible solutions. Preliminary results from sectors such as banks, universities, army headquarters, including metered energy data and analysis of user's behaviour are given in the present paper. Moreover, savings measured with available Energy Star compliant devices, are calculated and compared to previous estimates. Power quality, an increasingly concern for both building and electric utilities managers is also briefly discussed. Finally, recommendations are made aiming to improve, in such a continuously growing key market, the dissemination of energy-

[1] The authors wish to acknowledge with thanks the provision of preliminary data by A. de ALMEIDA (University of Coimbra - Portugal), D. Beyrand (CENBG - University of Bordeaux I - France), M. Jensen (DEFU - Denmark) as submitted by A. ANGLADE et al. at the International Seminar on Energy Efficiency in Office Equipment and Consumers Electronics (Stockholm - Sweden Sept. 1996) (see NUT 96)

S. Nørgård et al. (eds.), Development with Sustainable Use of Electricity, 127-154.

efficient technologies. A global issue, namely the reduction of CO_2 emissions, is underlined as one of the most promising ways in such direction.

1. Introduction

Information and Communication Technologies (I&CTs) are now recognised as one of the fastest growing electricity end-uses in old and new industrialised countries. Since the early 90s, such awareness has made many Energy Conservation and/or Environment Protection Agencies very keen to actively control, in close co-operation with leading manufacturers, the resulting load growth taking measures to establish an efficient power management. A perfect example is seen with the US-EPA *Energy Star* Office Equipment [2] (E*) labelling programme, now characterised by world-wide acceptance and availability of products. More stringent programmes with similar goals, are supported by a few European countries, namely Sweden and Switzerland, within their Energy 2000 Programmes (ROT 94), and cannot be ignored. Moreover, an analysis of the self electricity consumption of such «efficient» I&CTs shows that the list of published material is not very large (ROT 93), (BEN94), (McM 95), (MOH 96). New data, concerning similar issues in the «Brown Goods» sector were recently discussed at the International Seminar on Energy Efficiency in Office Equipment and Consumers Electronics (Sept. 1996 - Stockholm/Sweden) co-sponsored by EU DG XVII and Nutek, the Swedish Board for Industrial Development (NUT 96). There are many good reasons, listed hereafter, to carry out a more accurate assessment and analysis of related issues. First, most available data are issued from manufacturer's declarations or specification sheets. Second, most buildings used in tertiary activities are designed or operated, in such ways that the consumption of electricity by office equipment is underestimated or fully ignored. Third, building managers are becoming more and more interested in a fair knowledge of the money savings potential on the electricity bill. Fourth, the degradation of power quality due to the larger and larger amount of electronics circuits and related AC-DC converters in home electronics and business appliances may become a very important concern for electric utilities. Finally, we are absolutely convinced that, all actors involved in the office equipment sector when shown the impact of I&CTs on power load and electricity consumption agree upon the need for more and better knowledge of facts and factors. All they need is better information.

2. The MACEBUR [3] Project

The MACEBUR project was funded in 1995, within the EU-DGXVII SAVE/PACE Programme in order to analyse the actual energy efficiency of E* (or similar stand-by

---

[2] in the following, the acronym E* is used in replacement of US-EPA Energy Star Office Equipment labeling program

[3] Maîtrise des Consommations d'Electricité des équipements Bureautiques (Electricity Conservation in the office equipment sector).

modes) devices. Involving Danish/French/Portuguese scientists, the members of the task force are DEFU (Danish Board of Electric Utilities), Ademe (French Environment and Energy Management Agency) and researchers from Universities of Bordeaux (France) and Coimbra (Portugal). Basically, the work will include: (i) a campaign of measurements, (ii) an approach based on a study of user's behaviour and of the attention given to these issues by large purchasers and commercial engineers through interview, questionnaires, ... (iii) information campaigns aimed at users and decision-makers, including buyers, mainly in order to improve their overall knowledge. The following data for which only preliminary results are presently available, are particularly interesting :

- the actual power load of recently purchased office equipment, which perfectly demonstrates user's level of ignorance and/or support of the energy-efficient [4] of their own equipment,
- the present state-of-the-art in recently built or retrofitted or more ancient office buildings in terms of direct and indirect (such as air-conditioning) power load and consumption,
- a feasibility study to implement a possible synergy approach including ergonomics and power quality.

In all three countries, the campaign of measurement started in the Spring of 1995 in different office buildings : army, banks, city halls, government, hospital, research laboratories, etc. . . About 600 sets of equipment and some 10,000 m² are presently being monitored. Some buildings are new, others more ancient and/or recently retrofitted. We propose to emphasise different parameters such as the relationship between heat loss by office equipment and other end-uses for example ventilation, heating and cooling, or the type of building and/or the activity. Moreover, the study also aims to identify links between energy efficiency and such concerns as ergonomics, the environment or power quality, to be possibly established in a near future by users or building managers.

3. Case studies

The following are preliminary results from a selection of case studies forming part of the MACEBUR project.

---

[4] In the present paper, the words «energy-efficient device» and «stand-by mode», used in the same meaning, indicate that the equipment shifts to a low power load when unused after a idleness period of time normally chosen by user itself. Up to now, this is the common situation. That may change in the future if more smart systems are designed.

3.1. BANKS IN DENMARK

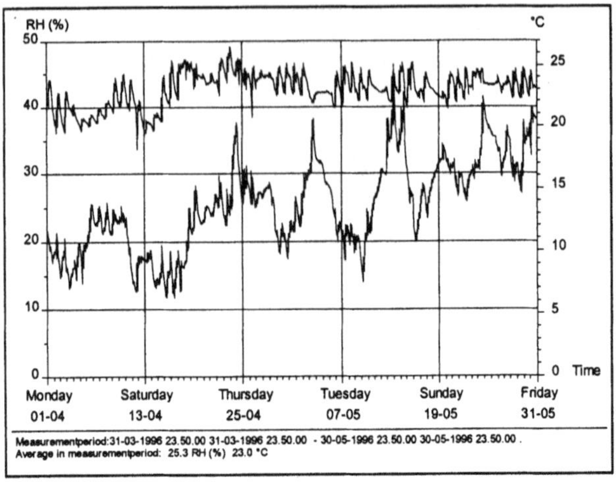

Figure 1.
Measured temperature (upper curve) and relative humidity (lower curve)
(JB headquarters in Zealand)

Jyske Bank (JB : Bank of Jutland), the fourth largest bank in Denmark and the most popular bank in Jutland, has offices in the whole country. Its regional headquarters on Zealand, chosen for the campaign, are located in an ancient, renovated 5 storey building. Local customer services are provided on the ground floor. The first floor, devoted to Private Banking International, was the measured site. A number of other banking or internal management services are operated on the upper floors. The main reason for JB's interest in this measurement was the unsatisfactory operation of a centrally regulated ventilation system that did not provide satisfactory standards of temperature and humidity on all floors equally i.e. room temperature was too high in the first floor and too low on the top floor. The measurements are expected to help investigate possible solutions. Typical data for standard equipment are shown in Figures 1 to 7. Most results were obtained over a period of approximately 1.5 month campaign (3 months in one case), from an observation of typical office equipment : PCs (including monitors), printers, fax machines and one copier. In order to achieve a complete analysis, different data have to be collected such as relative humidity and temperature in the rooms or lighting power loads (it should be noted that electricity consumption results mainly from lighting and office equipment). In Figure 1, the temperature and relative humidity data, recorded in April and May 1996, may help to understand why JB chose to reduce heat emissions from lighting and office equipment, the room temperature being relatively high during business hours : 23,4 °C in April and 23,6°C in May, on average. From measurements performed on 13 out of 52 PCs in 1st floor, we show in Figure 2, the average pattern of use during a business day for a typical PC system, including the monitor. The results unambiguously evince JB's policy for such equipment : while the machines are used intensively throughout working hours, they are turned-off at the end of day. In some cases, the measurements also show that, during lunch break, the monitors are automatically turned into the suspend mode. Several reasons may explain the fluctuations shown in Figure 2 : the brilliance of the display and, in the lowest values, a switch to a screen-saver mode or the voltage fluctuation of the grid. Figure 3 presents the energy consumption by one PC, (including the monitor) on a typical working day, showing that PCs and monitors, are very intensively used throughout the day. In such conditions, no decrease in electricity consumption is possible and no savings or decrease in heat dissipation can be expected.

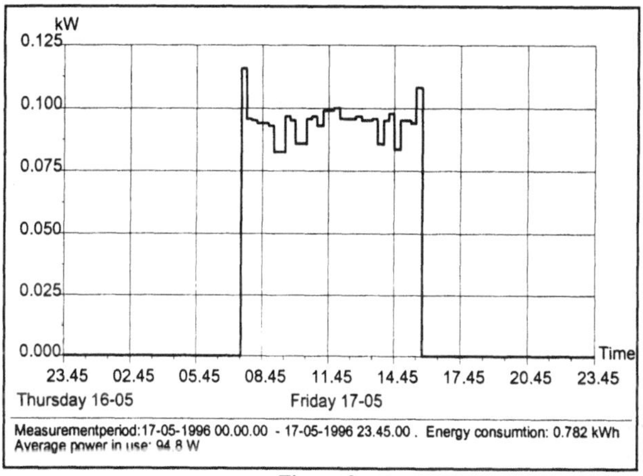

Figure 2.
Average, generated from about 45 days, PC daily use and energy consumption (JB headquarters in Zealand)

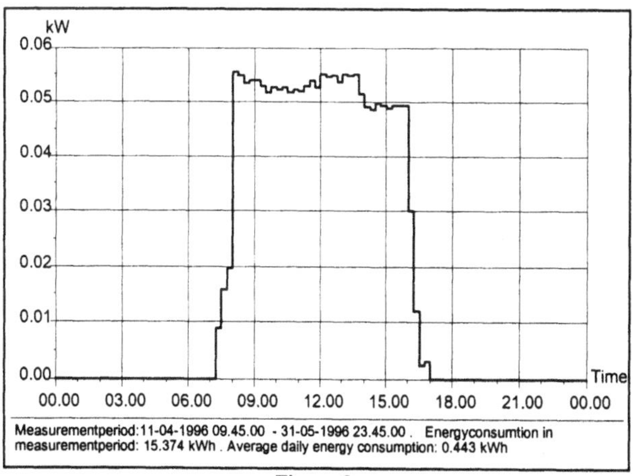

Figure 3.
Average daily PC electricity consumption (generated from about 45 days) (JB headquarters in Zealand)

Measurements were performed on 5 out of 10 printers on that particular floor. Figure 4 shows the average energy consumption of one printer (HP Laserjet IV) calculated for a typical working day over the same 1.5 month period (April-May 96). Intensive use of the printer during working hours and a reduced use in off-business hours are evidenced. As checked against energy consumption over weekends, the stand-by power load of that printer was 27 W. Reducing this to about 5 W would have brought the electricity consumption over the measurement period down to 30.4 kWh instead of 55.3 kWh, an energy reduction of about 0.6 kWh per printer and per day i .e..45 %. Over a period of one year, the reduction would therefore be 2 MWh for the series of 10 printers.

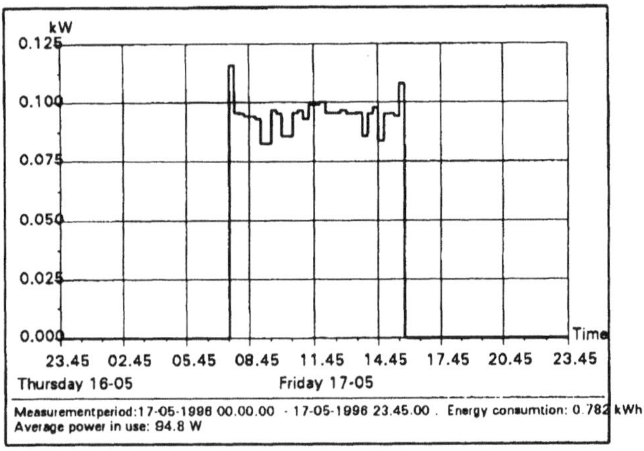

Figure 4
Average printer daily energy consumption (generated from about 45 days)
(JB headquarters in Zealand)

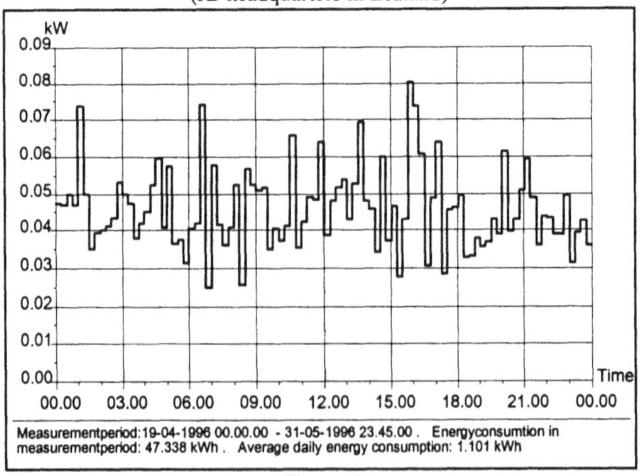

Figure 5
Average daily energy consumption of a facsimile machine (generated from about 45 days)
(JB headquarters in Zealand)

Figure 5 gives similar data for a facsimile machine (Canon L770) with characteristic ample fluctuations of the power load. Moreover, no difference was observed in energy consumption when the fax machine was either in the stand-by mode or in active-use, meaning that energy consumption remains constant whether or not the facsimile function is being used ; this is easily understandable given that the drum in such machine is permanently heated. A second fax machine with better specifications (including its being able to turn off in the absence of incoming or outgoing transmissions) was also measured : its average daily electricity consumption was 104 Wh, an estimated saving potential of about 90 %, obviously reducing the heat emission by the same percentage.

In OECD countries, as demonstrated in a recent IEA study (McM 95), copiers are now regarded as an important target for energy efficiency : out of 80 TWh for the whole office equipment sector, their consumption is evaluated to 13.1 TWh [5]. It is well known that the stand-by function in copiers, when installed, cannot be disabled [6]. However, it should not be ignored that copiers are an important source of heat emission. For that reason too, their power management, taking into account affordable users's requirements should be recommended.. Figure 6 shows the average daily use and energy consumption from data recorded over about 3 months : for unclear reasons the copier was not turned off systematically at night, only about every two days. The calculated night/week-end power load was about 216 W. Should this stand-by load be reduced from 216 to 50 W, and the copier turned-off at the end of every business day, the average daily energy consumption would be decreased by a factor 3, from 4.2 kWh to 1.4 kWh ! Moreover, when the copier in the "OFF" mode, the power consumption is as high as 10 W : if, as requested by the Swiss Target Value Programme (AEB 94), a lower "OFF" mode power load were available i.e. 3 W, higher savings would be obtained, the average daily energy consumption then being 1.3 kWh, a low but non-negligible additional decrease.

---

[5] Such figure, established from the Demand-Side only, does not include the losses on the grid or in building's transformers, nor the low Power Quality (see § 4.2) effects

[6] in some cases, this seems to have been implemented several years ago in order to avoid a long delay before the user could obtain the copy, this delay resulting in customer's complaints.

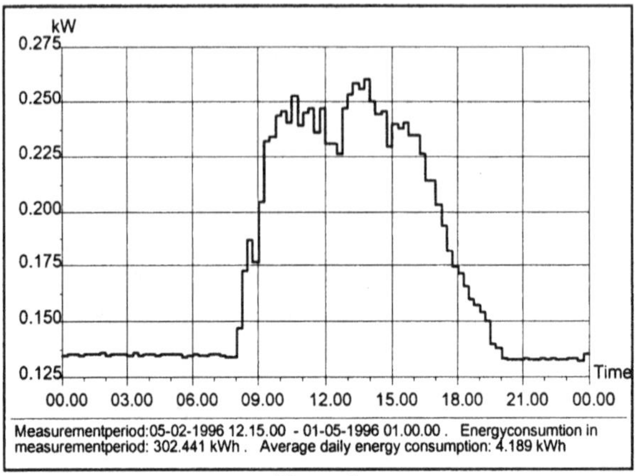

Figure 6.
Average daily energy consumption (generated from about 3 months measurements)
(JB headquarters in Zealand)

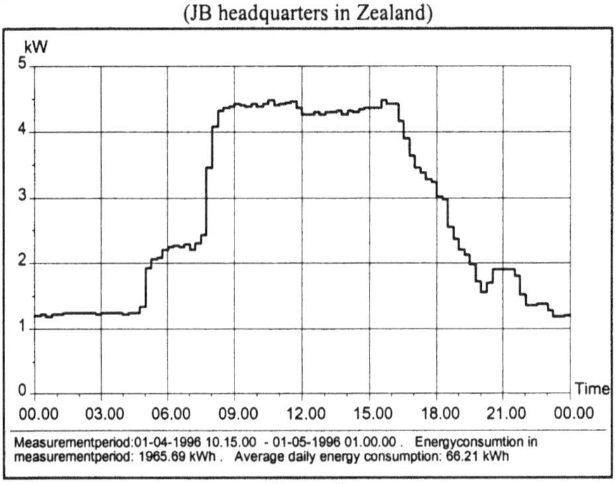

Figure 7.
Average (generated from 1 month measurements) daily lighting energy consumption
(JB headquarters in Zealand)

In the same place, the lighting power load was also metered enabling us to underline two main results. During off-business hours, the power load is typically 1 kW, all luminaries not being systematically turned off at night. Also, the data differ only slightly from January to March. As shown in Figure 7, it is obvious and well-known that, in such office buildings, lighting is another major source of heat dissipation. It is therefore possible to identify several solutions to decrease the electricity bill through energy savings from office equipment, lighting and ventilation systems. However, in such place, the most influential measure is to optimise both room temperature and relative humidity. Undoubtedly by improving the comfort of members of staff one can also hope to increase their productivity, a key argument in favour of greater energy efficiency.

3.2 UNIVERSITY OF COIMBRA (PORTUGAL): DEPARTMENT OF COMPUTER ENGINEERING

The Department of Computer Engineering is a 3-storey building that includes classrooms, computer rooms, teacher's offices, a cafeteria and a library. The total area is 4400 m^2. In this building, the stock of equipment comprises 94 computers (IBM-PC compatible and Macintosh), 12 workstations, 1 network server, 13 printers, 2 medium-size copiers. In this building, the study done during six days in mid-1995, has shown an electricity consumption amounting to 3.5 MWh. Extrapolated to the complete year, the consumption of this building is close to 200 MWh distributed as shown in Figure 8. Lighting is responsible for 54% of the total energy consumption, a very high figure resulting from the large number of lamps and bad lighting control, many luminaries remaining ON in the circulation areas more than necessary. Actually, as also shown in the Danish study, these field measurements may result in a possibly important by-product through the implementation of better lighting control producing non-negligible money savings. Office equipment in Coimbra accounted for 28% of the total energy consumption, of which about half for microcomputers and workstations.

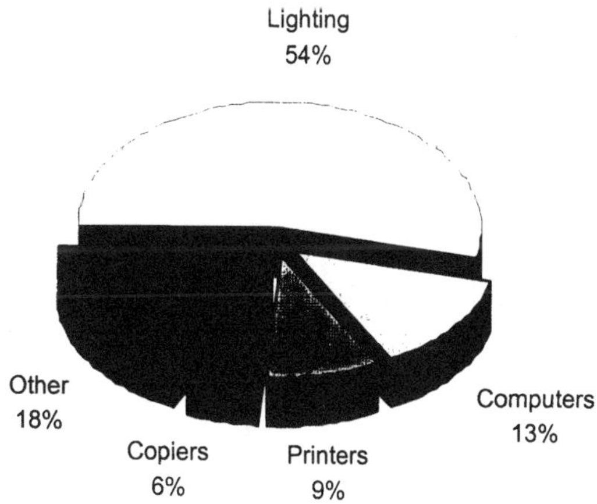

Figure 8.
Energy consumption by type of equipment
Department of Computer Engineering - University of Coimbra, Portugal

Figures 9 and 10, respectively, present the load diagram and the power factor in the Coimbra building over one week. The solid and dotted vertical lines respectively refer to 0 AM and 12 AM in the same day. In such a Research Centre the machines are obviously used 24h a day. When compared to JB's data, an obvious difference appears, i.e. the relatively high consumption and resulting bad power factor, during the night and over weekends. In such cases, the power factor due to the high penetration of compact fluorescent lamps and office equipment use, both having a very poor power factor (PF) in a typical 0.5-0.7 range, vary very deeply. The current waveforms in the 3-phases as shown in figure 11-a, -b, -c, are distorted. This results in very heavy harmonic pollution, a possible source of misoperation in particular in breakers and other safety equipment not designed to take such waveforms into account.

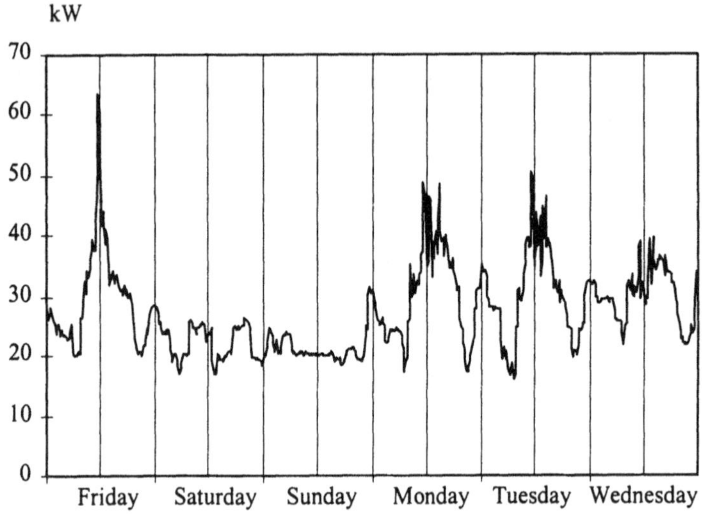

Figure 9.
Load diagram of the Computer Engineering building (6 days in mid-95)
Department of Computer Engineering - University of Coimbra, Portugal

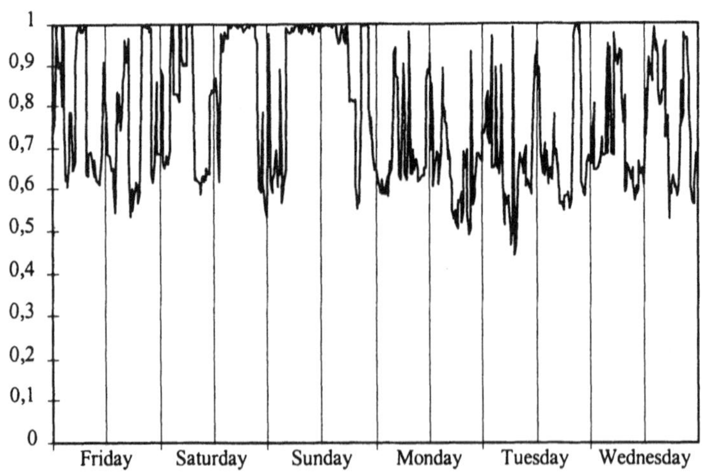

Figure 10.
Daily evolution of power factor
Department of Computer Engineering - University of Coimbra, Portugal

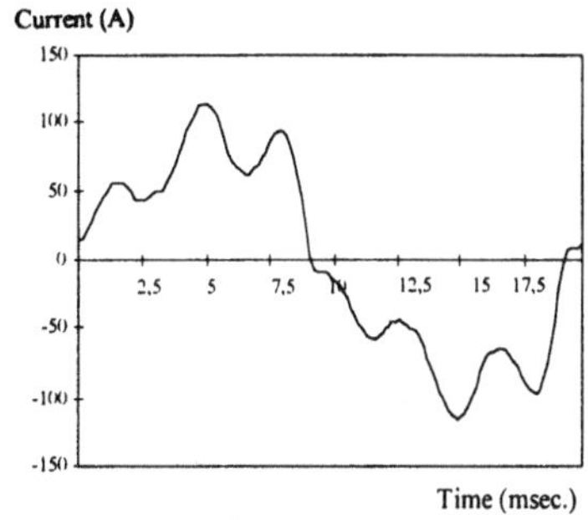

Figure 11-a.
Current waveform phases 1,2,3

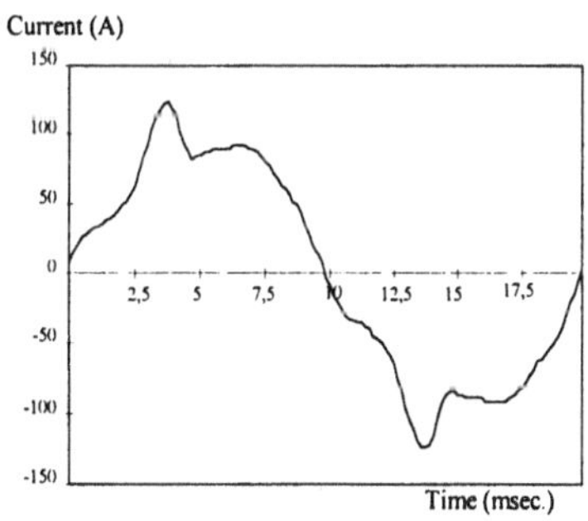

Figure 11-b.
(Dep. Computer Engineering)

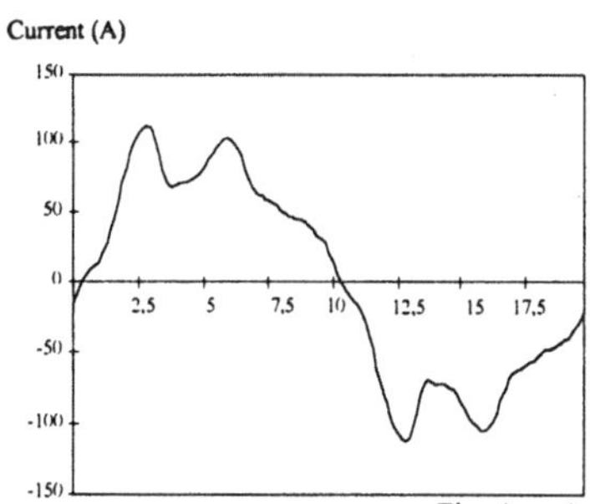

Figure 11-c.
(Un. Of Coimbra, Portugal)

3.3. FRENCH SITES

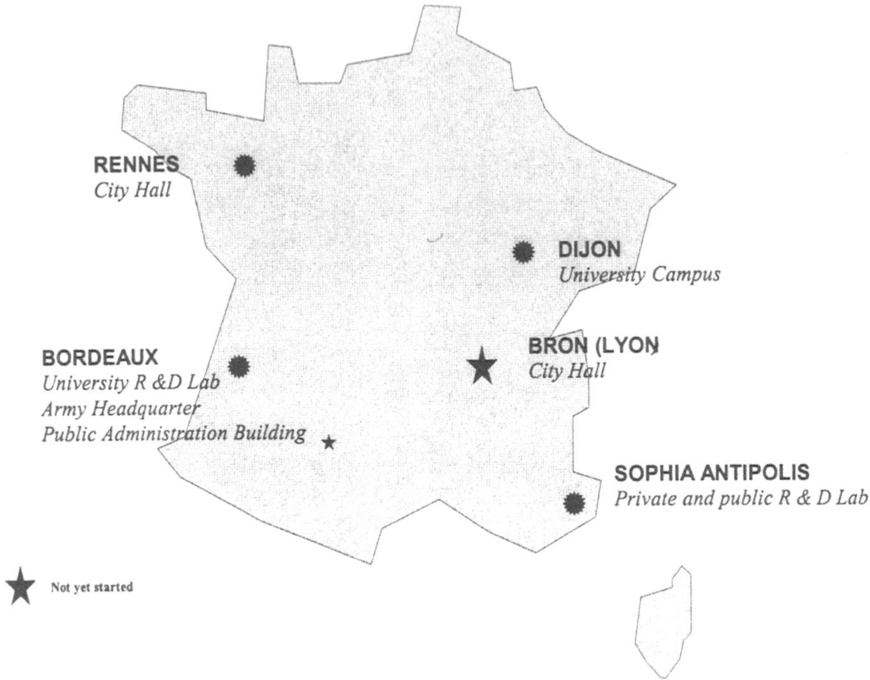

Figure 12.
Map of the locations of French sites

As shown in Figure 12, the French data were obtained in about 8 different companies from various sectors located in 5 areas : Ademe headquarters (Sophia Antipolis), Army headquarters (Bordeaux), City hall (Bron and Rennes), Research Laboratories (Bordeaux and Sophia Antipolis), University Campus (Dijon), Consultant (Sophia Antipolis). For the purpose of the present measurement campaign, a new type of handy wattmeter was developed in partnership by Ademe, CENBG and a Swiss manufacturer [7], the data being stored in a 256 Kbytes memory. Later read with an optical device, these data were logged into a portable PC (BEY 95). Fifty such wattmeters were typically used in a 3-week standard campaign with 7 physical data per record, stored every ten minutes. Data processing and analysis, using conventional computers, was performed later on, in CENBG. Preliminary results concerning the measurement campaign, are given here for 3 sites : Army Headquarters (Site # 1), a Research Laboratory in Bordeaux (CENBG : Site # 2) and a Municipal building in the City of Rennes (Site # 3). Table 1 gives the characteristic specifications of all main buildings : size, type, use, number of occupants ... Thanks to such metering of over 300 I&CTs pieces of equipment, a very

[7] EMU Elektronik AG - Erlibergstrasse 11 a - CH 6314 UNTERAGERI

large collection of data was obtained, enabling us to accurately compare all sites with one another, to identify the way equipment is actually operated and to suggest some realistic extrapolation. Figure 13 shows the I&CTs impact on the whole electricity consumption for every site.

The survey carried out in Site # 1 produced several interesting results. In that place, many items of E* equipment enabled during a first and preliminary campaign, 1 year before the present work, were found to have been disabled when we have started out our second campaign. Secondly, out of 30 % of E* compliant hardware, only 40 % was enabled. Such situations occur very often, users and computer managers initially not caring or ignoring the direct and indirect benefits of power management. However, we can easily demonstrate the benefits to them if only by showing Figures 14 and 15, which, in a "before and after" style, emphasise the impact of energy savings devices on electricity consumption. To-day, as a non-negligible by-product of the field measurements, many people become much more aware of the advantages of better controlled use - they are now willing to enable E* equipment, not satisfying themselves with the display's screen saver capability. Also, as a first step, users were advised to switch off their machines at night, this decision having a large impact on the electricity bill. Our analysis demonstrating the feasibility of savings up to 40 to 50 %, further steps should follow the present first one. Regarding printers, most of them having a stand-by mode but no ON/OFF switches, our measurements show a very large electricity leakage up to, about 98 % of the electricity consumption in extreme cases.

Table 1. Specifications of French Sites

Site	1	2	3
Office area (m²)	14 101	2 230	4 600
Employee's number	874	60	228
1995 Electricity consumption (MWh)	1 065	220	258
I&CTs equipment number	1 923	120	150
Metered equipment	50 + 41	51 +53	50 + 50
E* compliant equipment	577	n.a.	45
E* enabled equipment	260	n.a.	2
m²/employee	16, 1	38,8	20,2
Equipment number/employee	2,2	2,0	0,7
1995 Electricity consumption (kWh/m²)	76	94	56
1995 Electricity consumption (kWh)/employe	1 219	3 667	1,133
1995 Electricity consumption ratio : I&CTs /Total	9.8 %	23.8 %	22.9 %

n.a : non available

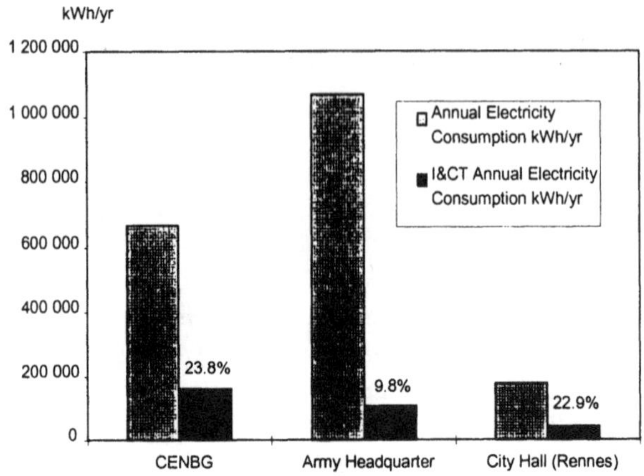

Figure 13.
Total and I&CTs annual electricity consumption in 3 French sites

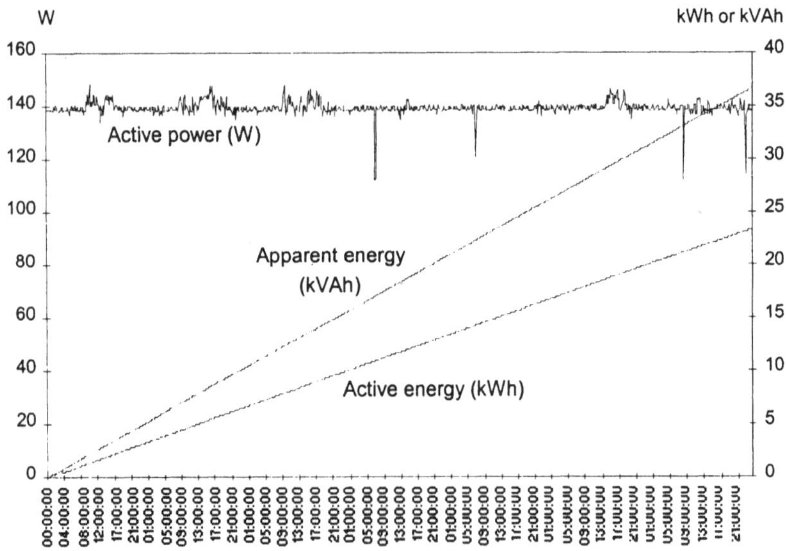

Figure 14.
21'' monitor : electricity consumption with stand-by mode disabled

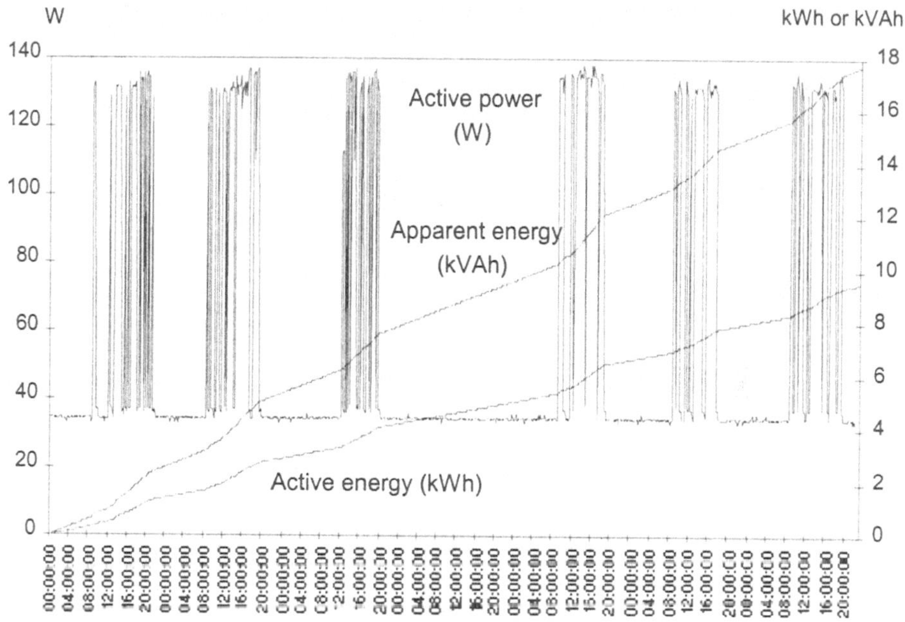

Figure 15.
21'' monitor: electricity consumption with stand-by mode enabled

4. Benefits/Obstacles

4.1. MONEY SAVINGS

From data obtained in Site # 2, the usage patterns and resulting electricity consumption of two 14" Monitors (E* compliant and no-compliant respectively), are shown in Figures 16-a, -b, -c. Electricity and money savings are calculated too. Taking into account possible further changes in energy policies, we also simulate an Ecotax extra-cost, the cost of electricity and Ecotax being assumed to be 8 and 1.5 US cents/kWh respectively.

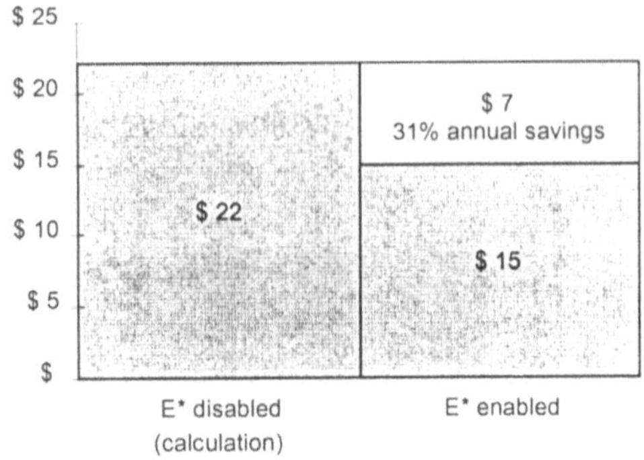

Figure 16-a.
Annual electricity bill for each (disabled/enabled) 14''monitor

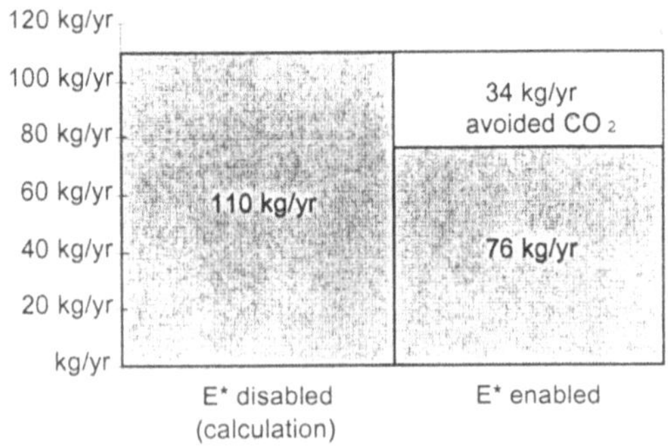

Figure 16-b.
CO_2 emissions for each disabled/enabled 14'' monitor

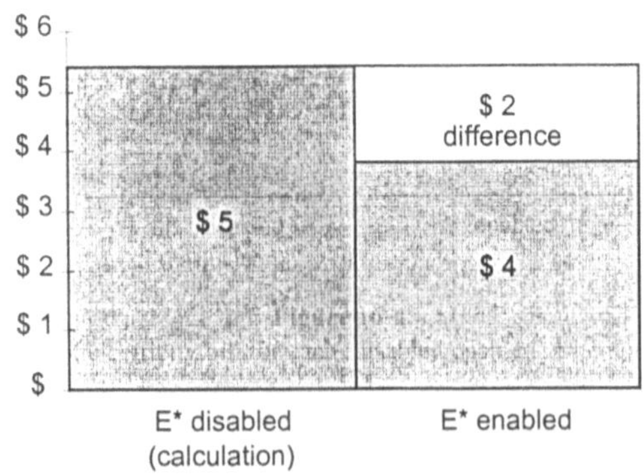

Figure 16-c.
Simulation of a possible Ecotax for each disabled/enabled 14'' monitor

4.2. POWER QUALITY

In §3.2, the Power Factor (PF) is identified as a possible source of problems (see also ALM 94) : typically 0.6 in normal mode in most equipment, it can collapse to a very low 0.2 in some printers and copiers turned down in stand-by mode, with nobody complaining about such a situation [8] . Actually the typical purchaser of a computer, mainly interested in the fastest and the cheapest equipment, ignores the amount of electricity consumption and does not want to be bothered with specifications regarding power quality. Only those who are engineers or concerned managers, may know that harmonic pollution occurs as a consequence of AC-DC converters systematically using switching power supplies either in office equipment or home electronic appliances (TVs, VCRs, wireless telephones, answering machines, etc.). Although all disadvantages of a bad PF and high total harmonic distortion have not yet been fully assessed, a few examples of the influence of harmonic pollution may be identified :

- an increase of total electric intensity, leading to a higher thermal losses and possible resonance between inductive (lines) and capacitive elements,
- possible interference in communication lines, in particular line carrier systems
- a substantial increase of the current in the neutral of 3-phase systems, with heavy thermal losses, including possible overheating,
- a possible mechanical resonance inducing vibration in electric equipment, leading to a reduction of the expected lifetime.
- extra thermal losses in cables, transformers and lines both on the consumer and utility sides

[8] In a preliminary step of the campaign, a data logger using Power Line Carrier for data transmission, was found completely unable to measure a printer probably due to a very deep harmonic pollution generated by the printer itself

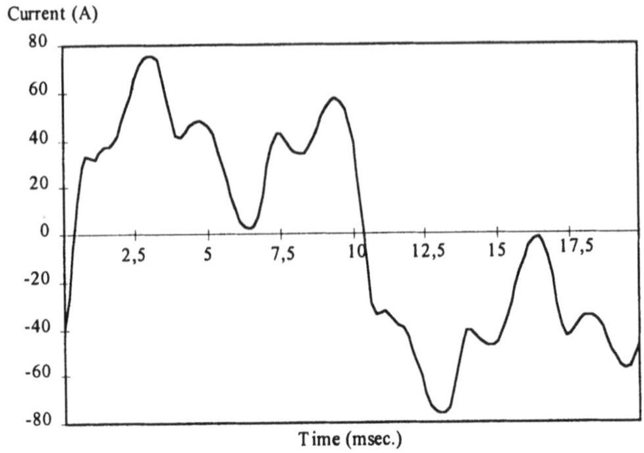

Figure 17.
Current wave form in the neutral wire during one period: 20 ms
Department of Computer Engineering - University of Coimbra, Portugal

However, up to now, such negative impacts of a low power factor are not frequently reported from the utility or the consumer point-of-view. Measured, in a transformer, on the neutral wire in the Computer Centre Building (University of Coimbra), a typical example of harmonic pollution is shown. Regarding the neutral conductor, the current waveform is shown Figure 17. Its RMS is about 60% of that of phase 1, 2 or 3. The relative harmonic intensity in the neutral conductor and the total harmonic distortion factor (A-THD) of the electric current are given in Table 2. Our present analysis is restricted to underline that when systematically enabling E* system, a smart way to limit the harmonic pollution and possibly ensuing misoperations, is made possible.

Table 2. Fundamental and harmonic currents in % of I_{rms} *Department of Computer Engineering - University of Coimbra, Portugal*

I_{rms} (A)	Fundamental and harmonic currents (% of I_{rms})								A-THD (%)
	Fund	H 3	H 5	H 7	H 9	H 11	H 13	H 15	
45,2	81,4	52,4	10,5	3,8	22,2	1,2	1,0	5,1	58,1

5. Present situation and future trends

Some trends emerge from the present preliminary results of the monitoring campaign, as in the data analysed below, regarding in particular both the user's behaviour and a company's policy.

5.1. OFFICE EQUIPMENT IN FRENCH OFFICE BUILDINGS

Statistical studies, regularly carried out on a yearly basis in France, aim to give a detailed view of main economy sectors and sub-sectors using accurate indicators. In correlation with the present study, one of these studies [9], was recently performed (late 1994) through a "snapshot" of the office equipment energy consumption in the office buildings sub-sector conducted by CEREN. The poll, performed on a sample of 1481 buildings was double-checked with national energy balance data, showing a low 4% range of error. From such results and data obtained from field measurements, possible trends and recommendations in terms of energy efficiency strategies were defined, as given in § 5.3. The main results from the 1994 CEREN study are as follows :

- in more than 99 % of the total investigated area, office buildings, are equipped with at least one I&CT system,
- the typology, or present computers installed base, represented in figure 18, shows the effect of downsizing,
- the ratio of workstations, that include PCs, connected to a server is 69%, greater than the 36% ratio of dumb terminals/workstations connected to a minicomputer/ mainframe.
- 45% of PCs are networked, some of them being both linked to a server and a mini/mainframe,
- 51% of PCs were bought later than January 1993. Taking into account the availability of E*, compliant machines on the French market, we assume that PCs bought before the end of 1993 were not E* compliant and most PCs bought from 1994 to now are,
- regarding the daily Time-of-Use basis, the more recent the PC, the longer the time used (Figure 19).

[9] Participants were Ademe and other national bodies: Institute for Economical and Statistical Studies (INSEE), the main energy utilities, including EdF the French electricity board, and the Centre for Economical Studies and Research in Energy (CEREN).

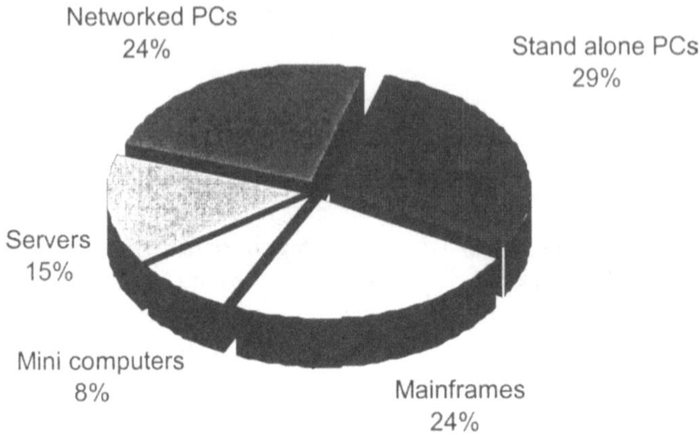

Figure 18.
Computers French installed base (1994)

The total French I&CTs electricity consumption was been assessed from those parameters, showing the strongest correlation coefficients such as the average number of monitors per building or the average area or the average electricity consumption. A convenient extrapolation enables us to calculate the I&CTs electricity consumption, 3.4 TWh [10] in 1994 for the office buildings sector. When compared to other energy end-uses, this is equivalent to :

- 7.8% of the total energy consumption, or
- 15.7% of the total electricity consumption,
- 22% of the non-thermal uses of electricity (other end-uses being lighting, motors, air-conditioning and ventilation ...)

[10] Based on the average European CO_2 emission intensity of 450g/kWh$_{el}$ produced, 3.4 TWh represent an annual emission of 1,5 MT of CO_2,

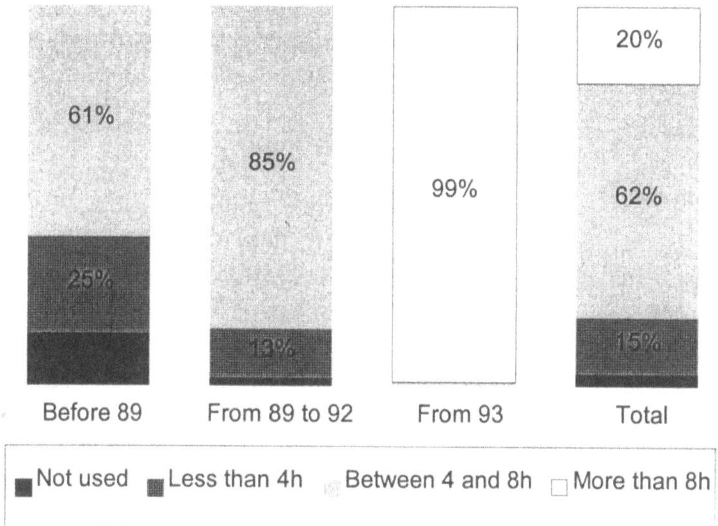

Figure 19.
PCs time-of-use - Influence of the year-of-purchase
(1994 CEREN poll on 1481 French office buildings)

5.2. VENDOR'S KNOW-HOW AND USER'S BEHAVIOUR

For more than 2 years, E* labelled equipment has been commonly supplied in Europe. However, their non-efficient use of electricity is demonstrated here. In particular, from an analysis of a few case-studies, we have shown that users, and vendors alike remain uninformed or unconcerned. Our analysis of the situation was performed in different ways, firstly through interviews carried out in computers stores (January 1995 : Paris/North-Eastern France areas and Luxembourg City). In its first part, the typical interview aimed to assess the vendor's opinion and/or know-how. Surprisingly, but not totally unexpectedy, it was found that nearly all vendors were totally unaware of E*/Nutek labels. In the very few cases when their knowledge was not so bad, they never suggest that the customers should pay any attention. More curiously, one of the vendors was found introducing E* as nothing but a very temporary fashion, soon to be forgotten ! When preparing the field measurements presented here, other interviews were conducted through meetings with users and/or managers. Nobody really aware of E* equipment use was found, the monitor's "Black Screen" or "Screen-Saver" mode being usually identified as the E* energy-efficient capability. Finally, over 3 years in a row beginning in 1994, contacts were established with vendors in a Computer's Exhibition in Paris,. The first year, vendor's information was quite low, then much better in 1995 ; and again, curiously, it was found in 1996 to be as poor as 2 years before ! This may partly explain, in each metered site, all computer engineers and nearly all users, have commonly ignored E* labels : as they have never been properly informed, they know absolutely nothing about the stand-by capability. Moreover, such lack of interest is sometimes regarded by vendors as sufficient to justify their own lack of concern ! The present not-too-preliminary market analysis, restricted to the French case-study, is in accordance with the Danish and Portuguese situation and, as far as we know, with the European market as a whole [11]. This is why, it may be appropriate to suggest to manufacturers that they supply their European customers (i) with E* enabled equipment, as requested since fall 1994 in the USA by the EPA's Memorandum of Understanding (ii) with basic information and training for the whole purchasing chain, from vendors to users. At the end of 1996, as we still do not have any evidence that significant changes are taking place, quite the contrary in fact, we continue to feel that an improved knowledge is necessary, in order, hopefully, to overcome the obstacles.

5.3. POSSIBLE CONSEQUENCES ON THE ACTUAL EFFECT OF E*PRODUCT

From the French preliminary results only, a few indicators have been calculated on an average basis for some E* compliant equipment. In terms of avoided energy consumption, it was thus possible to define E* equipment ''efficiency'':

- stand-by (including suspend when available) and normal modes represent 56% and 44% of total time-of-use respectively ;
- PCs and printers show an average power ratio of 29% between stand-by and normal modes;
- the Leaking Electricity ratio (LER : ratio of electricity consumption in normal mode/stand-by mode), typically about 50 %, may be higher than 90 % for printers or fax machines
- only 20% of E* PCs are actually enabled (this includes French Site # 1 where the information was started a year before field measurements). There may be several reasons for this situation e.g. (i) whereas it was shipped E* enabled by manufacturer, the equipment may have been reset locally by the company's computer staff (ii) in simultaneously used equipment (i. e. central unit versus the display and/or new add-on card), non-compliant saving modes may exist (iii) the set-up procedures may be very uneasy (iv) there may be no compliance between E* enabled networked equipment and workstations. However, it is to be hoped that such misoperation factors will be progressively reduced, solutions being under study.
- regarding the 1994 PCs annual market with 51% E* compliant machines, the calculated contribution of the latter to a decrease of the whole electricity consumption by PCs was only 4%. For the whole I&CTs, the savings were limited to a poor 2.7%.

---

[11] As explained in the introduction, Sweden and Switzerland are countries where a very active policy is managed in that sector. However, to our knowledge, no data covering typical sites have been yet published showing a more energy-efficient PCs use.

Equipment with a stand-by mode capability, most of these E* compliant, is widely distributed in Europe ; some machines, however, are non-labelled because of a too high value of the stand-by power-load for example (see Figure 15). Due to the expected saturation of the market for E* products and the quick rate in equipment replacement, non E* products will be out of use (or out of market) before the end of the present decade. In such a situation, 100% equipment E* compliant being enabled, through the same time-of-use and power indicators, 40% savings are possible. More stringent specifications i.e. a decrease to 15 W in the present 30 W E* standard, could raise the savings to 48%, this obviously requesting, as previously noted, E* enabling equipment. Furthermore, were the downsizing trend to be confirmed too, similar savings would be achieved in the entire I&CTs sector.

Such calculations apply to the active energy (in Wh), the energy charged by the utility to the consumer and not to the apparent energy (in VA), the energy delivered by the power plant at the busbar. To our knowledge, no detailed final assessment of the actual electricity savings from the electric utilities point-of-view is yet available. However we cannot completely ignore this issue, possibly important in countries where the shortage of electricity is a permanent risk. For this reason, we would like to stress two important facts. First of all Power Quality, previously presented (§ 4.2), may be improved in new power supplies, through built-in Power Factor Correctors (PFCs), manufacturers announcing many direct benefits ;

- 40% savings in energy consumption compared to the present situation ;
- low extra-cost expected to decrease as PFCs become standard equipment ;
- with a no set-up /no tune operation, the user is not a "limiting factor" ;
- with a PF = 0.99, the energy consumption decrease applies to the apparent energy, almost equal to active energy.

However, we should not ignore possible hidden disadvantages for example in electric safety or the PFCs self power load. This may result in a more difficult step towards a decrease of stand-by power loads

The "OFF" consumption is a second important issue, evidenced particularly in Site # 3 (Rennes City Hall). In most cases, the power is used with a poor efficiency only to maintain in charge the batteries powering the configuration backup memories or clocks. Amounting to 2% of the total I&CTs consumption this is not a ridiculous figure when compared to the calculated possible E* benefits (2.7% in 1994). Ideally, while quite unrealistic as yet, the equivalent power load may be cancelled by a manual plug-off from the wall socket. Anyway, in spite of existing and possibly not-too-expensive solutions in regards of the present market, the manufacturers have no incentive to implement any smarter technique ; this will be so as long as more stringent labels are not defined and implemented.

6. Comparison with other data

The cost of the field measurements being very high, a limited number of papers or reports only is available to date. Some national or regional surveys were released recently in Stockholm (NUT96, a Conference co-organised by Nutek and the European Union. The UK office buildings case-study has been presented by P. HILL (HIL 96) showing that « ... Office equipment is second only to heating as the major user of energy on offices ... ». Its electricity consumption doubles every few years, accounting to-day for about £300 M ($ 450 M) annually. Out of all end-uses, that sector is « ... perhaps the easiest for energy savings ... ».

Two other studies are also briefly summarised below. The first was performed by US scientists. The second being, as far as we know, is the very first and unique case-study ever performed in a new industrialised country, namely Thailand.

6.1. ENERGY SAVINGS AND PERFORMANCE OF POWER-MANAGED PCs AND MONITORS IN US

The study was started in 1994 by scientists from the Lawrence Berkeley National Laboratory (LBNL). A survey of 500 users of LBNL revealed that nearly half left the equipment ON 24/24h. Obtained from a significant sample, the results (NOR 95) confirm our data : for example, one of the most significant results is the fact that a systematic audit of several dozen compliant devices showed only 11 % PCs (plus 41 % partly) and 39 % monitors were E* enabled. In the said study, savings were estimated through three scenarios [2]: « As-Operated» (actual use), « Standardised » (typical operating pattern) and « Maximum » (achievable savings). In the « As-Operated » scenario, savings range from 0 to 75 kWh/yr possibly growing to 200 kWh/yr. mainly owing to improved monitor power management. One obvious conclusion is that, in US where the information is abundant and perfectly supported by many bodies, end-users have nevertheless not become very much more conscious and active « players » than their European colleagues. Such findings may confirm that a very long path remains yet to be travelled before the 1992 US-EPA's objectives become a reality. A dramatic improvement, i.e. through a *per default* entirely automatic stand-by mode, deserves to be included in the E* MOU !

6.2. THE THAI CASE-STUDY

To our knowledge, Thailand is the only non-OECD country where such analysis has ever been done (MOH 96) [14]. In this South Eastern Asia country, with a permanently hot, often very humid, climate, the demand for Air Conditioning (AC) is very high, frequently representing up to 60 % of the electricity consumption in commercial buildings. Moreover with a 8-10 % economic growth for several years, the increase in annual power demand has typically been 1 GW, a growing part of this being due to office equipment (from the base-case [1993] to 2005, a huge I&CTs market growth is expected i.e. 5 Millions PCs, 7 Millions monitors, 890 000 copiers, etc.). Contributing both directly and indirectly to the power demand growth, the office equipment end-use sector was very carefully checked, some important results being shown below:

- idle losses are quite high, from 53 % for a PC to more than 90 % for copiers, printers and faxes. Office equipment direct consumption ranges presently in a relatively low 2.2-5.6 % of the total electricity consumption of which a non-negligible part can be saved at no extra-cost.
- from a very few GWh in 1993, in the whole commercial buildings sector, the annual consumption is expected to grow very rapidly to 2,500 GWh in 2005 (Base-Case scenario). It should be possibly limited to 1,800 GWh, through the Energy Efficiency scenario, by year 2005. The expected savings in consumption and power load are high, amounting to 700 GWh and 200 MW respectively.
- in Thailand too, such results may be obtained only through voluntary side-programmes i.e. consumer education, market pull strategies and/or mandatory standards or labels.

7. Conclusion

Particularly in Europe, and probably in other countries too, as far as we know, in those sectors concerned by the efficient use of electricity, decision and/or policy makers are not paying very great attention to the electricity impact of I&CTs. However, in a few European countries, specially Sweden and Switzerland, as well as in the USA, the concern for I&CTs, considered as a major sector of energy end-use has grown steadily over the last 4-5 years. However, in our opinion, a systematic search for improved solutions limiting the resulting power load or electricity consumption has not been carried out adequately, for a number of reasons. One of these may lie in the present I&CTs energy-efficiency programmes which are adapted from those concerning lighting or home electric appliances, for example with standards and labelling programmes defined and implemented with the participation of manufacturers. Although such step look necessary, we think that the special features of I&CTs have not, up to now, been into account systematically enough. While compact fluorescent bulbs or refrigerators/freezers, remain energy-efficient during their whole life, no user having any possibility to modify anything, that is not true for PCs. In our opinion, this issue should be taken into consideration much more seriously by manufacturers and software companies.

A second piece of evidence should not be ignored : for many years, computer designers have been convinced that any improvement in the Integrated Circuits Technology, aiming to lower the internal heat losses, in order to increase of the number of components per cubic centimetre, was the perfect single criterion to achieve energy efficiency. While it was commonly shared by manufacturers in early days, this view was dramatically changed when the US-EPA "Energy Star Computers" label became a "de facto" mandatory requirement, thereby introducing the need for real power management of the whole equipment. On the contrary, in most countries all electric utilities, very much aware of DSM programs in the early/mid 80s, are now often become much more concerned by their present over-capacity, and prefer to increase their electricity sales rather than optimise end-uses. As is well known, I&CTs in spite of a very high electricity conservation potential, are not frequently posted in the top part of the possibly existing list of end-use sectors targeted by DSM/IRP [12] programmes. However, the growth of non-linear loads such as those associated to I&CTs (office equipment or home electronics), and the resulting larger and larger harmonic pollution are so high that electric utilities are facing a new emerging concern, now requiring some specific attention. Systematically promoting energy-efficient I&CTs may bring about some benefits in that area too.

Up to now, the main goal of E*, namely a decrease by about a factor of 2 of the electricity demand in the office equipment sector has not yet been achieved. More than ever, major improvements to the present E* programme are desirable. A dramatic change may be foreseen in two directions. First, a very friendly mode of use has to come as soon as possible, the equipment hopefully having the capability to manage its own power load through an on-line device i.e. an ability to consume only the electricity requested by the application underway. Secondly, as underlined above, the huge growth of new microprocessor-based apparatus, not limited to business applications, is quickly exploding in homes as well : home electronics is another I&CTs sub-sector, in which the energy-efficiency also deserves a global approach ! A boost to more stringent E* specification is unavoidable, many reasons demonstrated here, possibly demonstrating its present limited success. This is easily shown by world-wide examples analysing both metered consumption and users/vendors/managers' behaviour. The very weak price signal should be completed with a second one retrieving its initial importance (as in 1992, when the US-EPA decided to implement the E* Programme, which resulted from a comparison between the decrease in CO_2 emissions thanks to the annually saved electricity and that generated by 8 millions of US cars). Announced, in a recent OECD-IEA study (MUL 96) initiated by the Experts Group on the Framework Convention on Climate Change (FCCC), such a signal should be clearly understood by all policy-makers and/or users in all countries. In the latter work; Products Standards are analysed through different scenarios, each of them being more or less a contribution to a decrease, in electricity consumption and the resulting emission of CO_2 (see footnote 10). Two families of products in widespread use all over the world, office equipment and

[12] DSM : Demand-Side Management - IRP : Integrated Resource Planning

refrigerators/freezers, were investigated.[13]. Three scenarios, including expected market growth and gradual use of more energy-efficient components or circuits, were defined : «*Basecase*» the present situation, «*Introductory*» the systematically enabled and operated E* label, «*More Rigorous* » based on the Swiss Target Value Programme (AEB 96). Three global regions were considered namely Japan, North-America and Western Europe. The results unsurprisingly show that in the «*Basecase*» scenario, CO_2 emission has been and will be continuously growing from 30 MT in 1995 to 45 MT in 2005. With the «*Introductory*» scenario, emission is stabilised while in the «*More Rigorous* » scenario, it is limited to about 17 MT. The latter scenario therefore seems to be the only one we should support : it is technically feasible, the perfect «Only Winners - No Looser» solution.

8. REFERENCES

AEB 94 B. AEBISCHER «THE STATE-OF-THE-ART : SWITZERLAND» in OT3E Final Report

AEB 96 B. AEBISCHER in (NUT 96)

ALM 94 A. de ALMEIDA in ROT 94

BEN 94 L. BEN ALI - LA NOTION DE CENTRALES ELECTRIQUES VIRTUELLES DANS LES TECHNOLOGIES DE L'INFORMATION : ÉTUDE BASÉE SUR QUELQUES IMMEUBLES IMPORTANTS DU SECTEUR TERTIAIRE - Thèse de Doctorat en Instrumentation et Mesures (Université Bordeaux I - Mars 1994)

BEY 95 D. BEYRAND « Compte-Rendu d'essais de validation d'un dispositif de mesures et d'enregistrement d'énergie et de puissance électrique » - Internal Report CENBG 9541 (Unpublished) University of Bordeaux 1

HIL 96 P. HILL in (NUT 96)

McM 95 J. McMAHON, M.A. PIETTE, J. KOLLAR «OFFICE EQUIPMENT : MARKET CHARACTERISTICS, ENERGY USE AND STANDARDS HARMONISATION in ROT 94 AIE WORKSHOP PARIS 1995

MOH 96 B. MOHANTY and W. MUNGWITIITKUL «Energy efficiency of Office Equipment in Commercial Buildings in Thailand » Asian Institute of Technology Bangkok - Unpublished Report Aug. 96

MOL 94 O. MOLINDER «THE STATE-OF-THE-ART : SWEDEN» in ROT 94

MUL 96 F. MULLINS «DEMAND-SIDE EFFICIENCY STANDARDS : PRODUCTS-Energy Efficiency Standards for Traded Products : Refrigerators/Freezers and Office Equipment » OECD Paris (France) Second Draft May 1996

NOR 96 B. NORDMAN, M..A. PIETTTE, K. KINNEY «MEASURED ENERGY SAVINGS AND PERFORMANCE OF POWER-MANAGED PERSONAL COMPUTERS AND MONITORS» in (NUT 96)

[13] In the present work, we restrict ourselves to the second category, one of us (J.R.) having been associated to the OECD/IEA study

NUT 96 International Seminar on Energy Efficiency in Office Equipment and Consumers Electronics (Stockholm - Sweden Sept. 1996)

ROT 93 J. ROTURIER and J.P. HARRIS «NEW LOADS IN OFFICE BUILDINGS : OPPORTUNITIES FOR IMPROVING EFFICIENCY OF OFFICE EQUIPMENT» in INTEGRATED RESOURCE PLANNING ed. by A. de ALMEIDA, A.H. ROSENFELD, J. ROTURIER,
J. NORGARD Kluwer Academic publisher 1994 #E 261

ROT 94 J. ROTURIER «ENERGY-EFFICIENT OFFICE TECHNOLOGIES IN EUROPE» OT3E Final Report (Commission of the European Communities Contract # EC-DG XVII 4.1031/E/92-07 (1994)

ENERGY-EFFICIENT MOTOR AND DRIVES

ANÍBAL T. DE ALMEIDA and PAULA FONSECA
Department of Electrical Engineering
University of Coimbra
3030 Coimbra, Portugal

Abstract

This paper presents an overview about energy-efficient motor technologies, namelly Energy Efficient Motors (EEMs), Variable Speed Drives (VSDs), efficient transmission systems, etc. It also identifies the main barriers to the penetration of those technologies as well as the strategies which can be used to overcome those barriers and contribute to make efficient purchase decisions, accelerating the penetration of energy efficient motor technologies.

1. Introduction

Electric motors consume over half of all electricity used in developed countries and typically 60-80% of the electricity which is used in the industrial sector. The use of energy-efficient motor technologies can produce substantial electricity savings which are also normally associated with process improvements and longer life of the motor drive components.

Motors are by far the most important type of load. They are used in all sectors and in a wide range of applications namely: fans, compressors, pumps, mills, winders, elevators, home appliances, etc. It is their wide use that makes motor drive systems one of the main targets to achieve significant energy savings. In order to reduce the losses of the power drive system, the optimisation of the motor drive system efficiency requires a careful assessment of the individual components of the drive system. To reduce these losses, numerous technologies have been developed, including Energy Efficient Motors (EEMs), Electronic Variable Speed Drives (VSDs), mechanical transmission systems, etc.

Motor systems improvements produce substantial electricity savings and bring up opportunities to save money, to improve plant efficiency and to reduce greenhouse

S. Nørgård et al. (eds.), Development with Sustainable Use of Electricity, 155-185.

emissions. Additionally, the profitability and competitiveness of many economic activities can be significantly improved through the efficient use of electricity, using energy-efficient motor technologies.

2. Key Factors Affecting Motor Systems Efficiency

The vast majority of the motors used in industry are squirrel-cage induction motors (Figure 1 and Figure 2) due to their low cost, high reliability and fairly high efficiency. There are no electrical connections to the rotor, which means that there are no brushes, commutator or slip rings to maintain and replace. The speed of an induction motor is essentially determined by the frequency of the power supply and by the number of poles of the motor. However the speed decreases a few percent when the motor goes from no-load to full load operation.

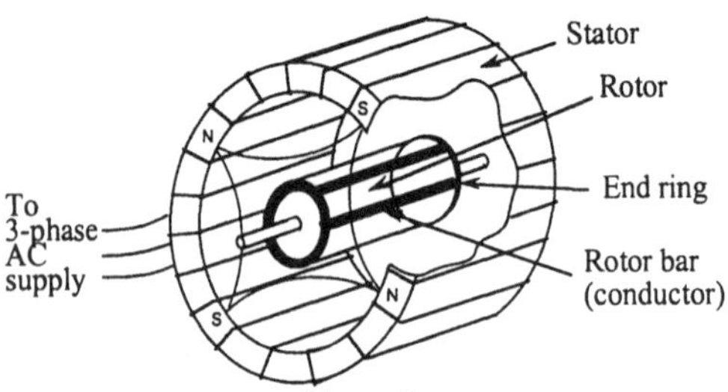

Figure 1.
Diagram of squirrel cage induction motor

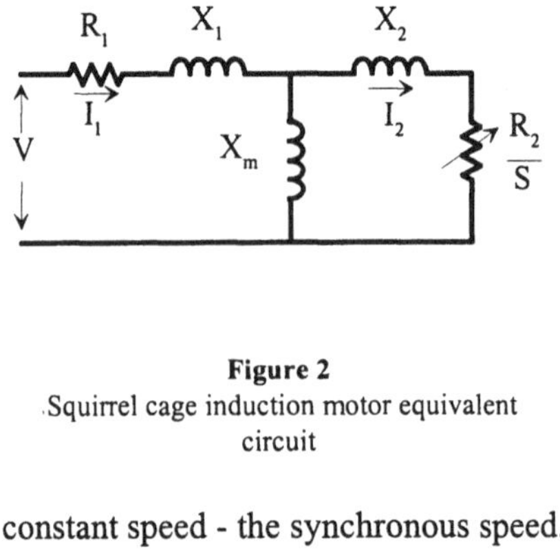

Figure 2
Squirrel cage induction motor equivalent circuit

Synchronous motors, rotate as name implies at a constant speed - the synchronous speed. The rotor of this type of motors is a wound rotor, which receives the excitation current from its excitation system. Synchronous motors, although they are more costly and present more maintenance problems than induction motors, are used in applications requiring constant speed, such as textile fiber industry and paper industry, high operating efficiency and controllable power factor. The last two factors are particularly important above 1000 kW. Synchronous motors, are also used in large plants to drive the air compressor. A large synchronous motor can be used to control the power factor of the whole plant, compensating the lagging power factor of a large number of medium and small induction motors.

The efficiency of a motor driven process depends upon several factors which may include:

- Motor efficiency
- Motor speed controls
- Power supply quality
- System oversizing
- Distribution network.
- Mechanical transmission
- Maintenance practices
- Load management and cycling
- Efficiency of the end-use device (e.g. fan, pump, etc.)

It must be emphasised that the design of the process itself can also influence to a large extent the global efficiency (units produced/kWh).

Process automation in industry can reduce labor intensity and increase productivity, as well as produce significant electricity savings per unit of output. Process issues tend to be application specific and are not in the scope of this paper.

2.1. MOTOR LOSSES

There are four different kinds of losses occurring in a motor: electrical losses, magnetic losses, mechanical losses and stray losses. These losses can be reduced by using quality materials, as well as by optimising the design. The electrical losses are of the type I^2R, and consequently they increase rapidly with the motor load. These Joule losses can be decreased by increasing the cross-section of the stator and rotor conductors. Magnetic losses occur in the steel laminations of the stator and rotor. They are due to histeresis and eddy currents, and vary with the flux-density and the frequency. They can be reduced by increasing the cross-section of the iron in the stator and rotor, by using thinner laminations, and improved magnetic materials. Mechanical losses are due to friction in the bearings, ventilation and windage losses. They can be decreased using low friction bearings and improved fan design. Stray losses are due to leakage flux, non-uniform current distribution, mechanical imperfections in the air gap, and irregularities in the air gap flux density. They can be reduced by optimal design and careful manufacturing.

2.2. ENERGY-EFFICIENT MOTORS (EEMS)

After World War II and until the early seventies, there was a trend to design inefficient motors which minimised the use of raw materials (namely copper, aluminium and silicon steel). These motors had lower initial costs and were more compact than previous generations of motors, but due to their low efficiency their running costs were higher. By the mid seventies electricity prices started escalating rapidly. Most of the large motor manufacturers started offering a line of energy-efficient motors besides standard efficiency motors. EEMs feature optimised design, more generous electrical and magnetic circuits and higher quality materials. Incremental efficiency improvements are still possible with the use of superior materials (e.g. amorphous silicon steel) and optimised computer aided design techniques. EEMs typically have 30-50% lower losses than the equivalent standard motors. Due to their lower resistance, EEMs normally have lower starting torque than standard motors. EEMs normally carry a price premium of around 20-25% in relation to standard motors. In a new application, and for a motor with a large number of operating hours, the paybacks are normally under two years.

In industrial plants most motors are bought incorporated in machinery built by original equipment manufacturers (OEMs). In order to decrease the equipment first cost, OEMs normally use standard efficiency motors. Therefore it is relevant for the plant operator to specify EEMs, namely when buying a large number of similar machines.

When a standard motor fails, and if it is a motor which operates a large number of hours per year, it presents a good opportunity for replacing it with an efficient motor. New installations also present good opportunities to install EEMs if they are supposed to run a large number of hours per year.

2.3. SPEED CONTROLS

Most induction AC motors are fixed speed. However, most important types of motor applications will benefit if the motor speed is adjusted to match the process requirements.

Motor speed controls are the devices which when properly applied can tap most of the potential energy savings in motor systems. Motor speed controls are particularly attractive in applications where there is a variable fluid flow. In many centrifugal pump, fan and compressor applications, the mechanical power grows roughly with the cube of the fluid flow. To move 80% of the nominal flow only half of the power is required. Centrifugal loads are therefore excellent candidates for motor speed control. Other loads which may benefit from the use of motor speed controls include conveyers, traction drives, winders, machine tools and robotics.

Conventional methods of flow control used inefficient throttling devices such as valves, dampers and vanes. These devices, although they have a low initial cost, introduce unacceptable running costs due to their inefficiency. Several speed control technologies can be used to improve motor system operation. Electronic Variable Speed Drives are discussed in detail because they are the dominant motor speed control technology. Figure 3 shows the power consumed by a motor driving a fan, using different flow control.

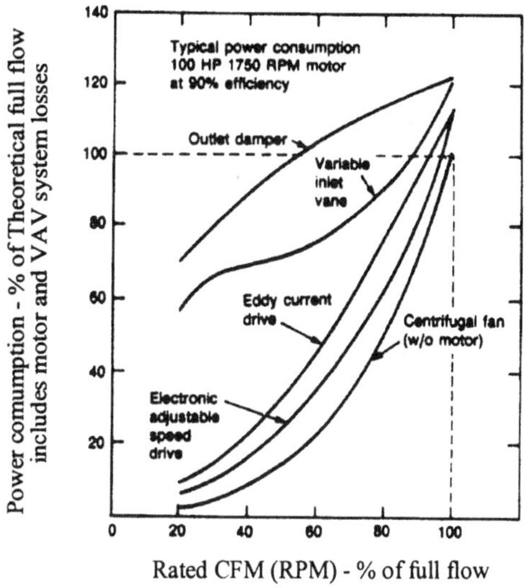

Figure 3.
Power consumption with different flow control methods in a fan system

2.3.1. Variable Speed Drives (VSDs)

Electronic VSDs are able to change continuously the speed of an AC motor. They work by simultaneous varying in a roughly proportional way the voltage and frequency of the power supplied to the motor. When the voltage applied to the motor varies linearly with frequency a constant maximum torque over the speed range is achieved (Figure 4). For loads with high starting torque or for very low speed applications, the voltage applied to the motor may need to be boosted to a compensate for the voltage drop in the motor resistance.

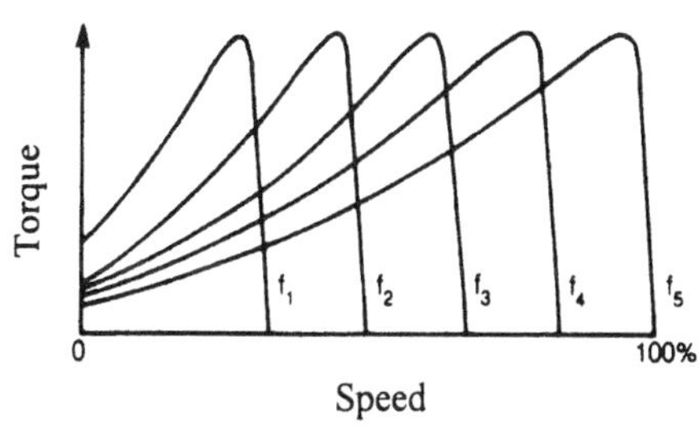

Figure 4.
Speed-torque curves for an induction motor (f1<f2<f3<f4<f5)

VSDs have no moving parts, presenting high reliability and efficiency. Because VSDs are not bulky and have flexible positioning requirements they are easy to retrofit.

The technology developments in the past two decades in the areas of microelectronics and power electronics made possible the design of efficient, compact and increasingly cost competitive electronic VSDs. As VSDs control the currents/voltages fed to the motor through the power semiconductor switches, it is possible to incorporate in the VSD motor protection features, soft-start and remote control, at a modest cost.

Presently there are VSDs for all types of motors available in the market. There are different types of VSDs, but the most common type is the inverter-based VSD. First the 3-phase supply is converted from AC to DC using a solid-state rectifier. Afterwards the inverter uses this DC supply to produce a 3-phase adjustable frequency, adjustable-voltage output which is applied to the stator windings of the motor. The speed of the motor will then change in proportion to the frequency of the power supply. Usually the output voltage waveforms can be synthesised over the frequency range of 0-100 Hz.

Figure 5 shows a general inverter power circuit with motor load:

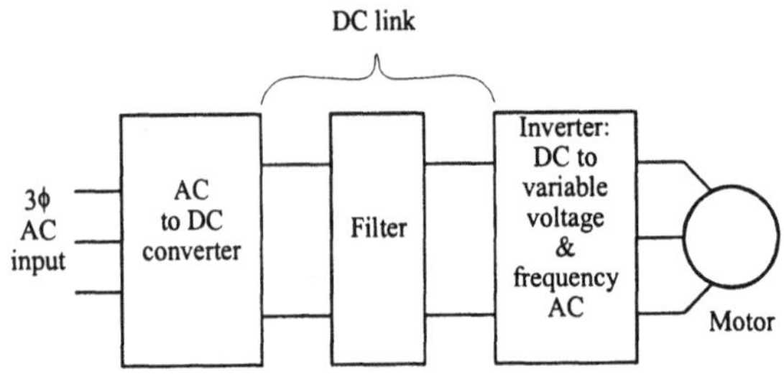

Figure 5.
General configuration of inverter based VSDs

The main benefits of adjusting the motor speed through the use of VSDs include better process operation, less wear in mechanical equipment, less noise, and significant energy savings (of 50% or more for some types of applications).

VSDs can have some disadvantages that should be considered. VSDs can produce electromagnetic interference (EMI) and can introduce harmonics into the supply leading to voltage distortion. To compensate the extra losses produced by the harmonics, some VSDs manufacturers, advise the reduction of motor rated power by about 10%. However the more effective way to mitigate these problems caused by harmonics is to install a modified designed motor, especially than EEM. Because EEMs have less losses, harmonics cause less over-heating and it is not necessary to reduce motor nominal power.

Table 1 summarises the advantages and disadvantages of different VSD drive types. There is not an VSD technology which emerges as a clear winner in all cases, when compared with other VSD types.

TABLE 1. Advantages and disadvantages of VSDs [1]

VSD Type	Advantages	Disadvantages
Pulse-Width Modulation (PWM)	Good power factor throughout speed range. Low distortion of motor current. Wide speed range (100:1). Multi motor capability.	No regeneration capability. Limited to VSDs bellow 1000 kW. Slightly (about 1%) less efficient than VSI or CSI
Six-step Voltage-Source Inverter (VSI)	Good efficiency. Simple circuit configuration. Wide speed range (10-200%). Multi-motor capability.	Poor power factor at low speeds (unless a rectifier/chopper AC/DC converter is used). No regeneration capability. Operation below 10% of rated speed can produce cogging.
Force Commutated Current-Source Inverter (CSI)	Simple and robust circuit design. Regenerative capability. Built-in short circuit protection. Wide speed range (10-150%).	Bulky. Poor power factor at low speed/load. Possible cogging below 10% of rated speed.
Load-Commutated Inverter (LCI)	Simple and inexpensive circuit design. Regeneration capability. Built-in short-circuit protection.	Poor power factor at low speed. Can only be used with synchronous motors.
Static Kramer Drive	VSD power is less than motor power. Can be retrofitted to wound rotor induction motor (W.R.I.M.) with external resistor.	Can only be used with W.R.I.M. Poor power factor at low speeds. Subsynchronous speed (50-100%)only.
Static Scherbius Drive	VSD power is less than motor power. Wider speed range (70-130%). Can be retrofitted to W.R.I.M. with external resistor if overspeed is possible.	More complex and costly than Kramer drive. Can only be used with W.R.I.M.
Cyclo-Converters	Can operate down to zero speed. High torque capability with field-oriented control. Can be used with induction and synchronous motors.	Cannot be used above 33% of input frequency. Complex circuit design. Poor power factor at low speed.

Pulse-width modulation (PWM) VSDs dominate in the low to medium power range (up to a few hundred kW) due to their lower cost and good overall performance. In situations when several motor drives must rotate at the same speed, those drives can be supplied by a single PWM VSD whose power rating is the sum of the motors ratings. In the range above a few hundred kW the choice of VSD technology depends on several factors including the type of motor, horsepower, speed range and control requirements. A few typical situations are:

- Large induction motors (above 1000 kW) with normal speed requirements for pumps and fans can use current source inverters.
- Large synchronous motors (above 1000 kW) use load commutated inverters.

- Large induction or synchronous motors (above 1000 kW) with very low-low speed requirements can use cyclo-converters.
- Wound-rotor induction motors (WRIM) can retrofitted cost-effectively with static Kramer or static Scherbius drives.

2.3.2. Multispeed and DC Motors

In applications where only a few operating speeds are required, multispeed motors may provide the most cost-effective solution. Multispeed motors are available with a variety of torque-speed characteristics (variable torque, constant torque and constant horsepower), able to cope with different types of loads. Two-winding motors can provide up to four speeds but they are normally bulkier (one frame size higher) than single-speed motors for the same horsepower rating. Pole-amplitude modulated (PAM) motors are single winding, two speed, squirrel cage induction motor that provide a wide range of speed ratios. Because it uses a single winding it has the same frame size of single speed motors for the same power, thus being easily retrofitted. PAM motors are available with a broad choice of speed combinations (even ratios close to unity), being especially suited and cost-effective for fans and pumps applications which can be met by a two-speed duty cycle.

2.4. SYSTEM CONSIDERATIONS

A number of important but often overlooked factors which may affect the overall motor system efficiency include: power supply quality (high quality power supply, careful attention to harmonics, etc.), system oversizing (proper equipment sizing), the distribution network that feeds the motor (attention to power factor and distribution losses), the transmission and mechanical components (optimised transmission systems), maintenance practices (careful maintenance of the entire drivepower system) and the match between the load and the motor (good load management practice), are discussed below.

2.4.1. Power Supply Quality

Electric motors, and in particular induction motors, are designed to operate with optimal performance, when fed by symmetrical 3-phase sinusoidal waveforms with the nominal voltage value. Deviations from these ideal conditions may cause significant deterioration of the motor efficiency and lifetime. Such deviations include:

Voltage Unbalance. Voltage unbalance wastes energy: it leads to high current unbalance which in turn leads to high losses. A phase unbalance of just 2% can increase losses by 25%. Additionally, long operation under unbalanced voltage can damage or destroy a motor (that is why many designers include phase unbalance and phase failure protection in motor starters). Another negative consequence of unbalance is the reduction of the motor torque.

Undervoltage or Overvoltage. When the motor is running at or nearly full load, voltage fluctuations exceeding 10% can decrease motor efficiency, power factor and lifetime.

Harmonics

Under ideal operating conditions, utilities supply pure sinusoidal waveforms (50 Hz frequency in Europe). However there are some loads, namely VSDs and other power electronic devices, arc furnaces, overloaded transformers, etc., that cause voltage distortion. The resulting distorted waveform contains a series of sine waves with frequencies that are multiples of the fundamental 50 Hz frequency, the so called harmonics.

Harmonics can increase the motor losses and noise, reduce torque, and cause torque pulsation and overheating. Vibration and heat can shorten the motor life, by damaging bearings and insulation. Harmonics can cause malfunctions in electronic equipment, including computers, induce errors in electric meters, produce radio frequency static and destroy power system components.

2.4.2. System Oversizing

The load factor probably is one of the most discussed factors on motor system efficiency. Designers usually tend to oversize motor systems to guarantee a growth in the system peak requirements. They assume that the extra capacity cost is a small premium to pay to insure the system will be able to cope with the maximum mechanical demand.

There are a number of reasonable reasons to install an oversized motor in some kind of applications, for example: if a process is a critical one, such as mixer in a batch control operation where the motor loss would mean a large financial loss, the decision maker may choose an oversized motor to insure against this possibility [2]. Oversized motors will allow high starting torque loads, they will be able to accommodate load fluctuations, and they will be able to operate under adverse conditions, such as voltage unbalance. If the motor is significantly oversized and operates at partial load, its efficiency may be reduced, meaning an increment in energy consumption, poor power factor and additional costs. As can be seen in Figure 6, the peak efficiency occurs normally around 75% full load for most motors except for the smallest motors, and above it does not vary significantly. In most motors the efficiency starts dropping significantly below 50% of full load. Manufacturers recommend the replacement of motors operating below 50% load by a smaller more efficient motor.

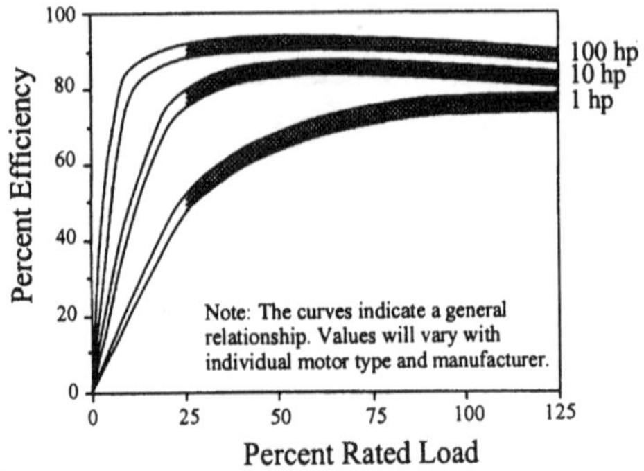

Figure 6.
Motor part load efficiency as percent of full-load efficiency [3]
for 1800-rpm three phase 60Hz design B squirrel cage induction motors

2.4.3. Distribution Network
There are substantial losses through the distribution network from the substation to the loads. These losses can be reduced by proper selection and operation of efficient transformers, by correctly sizing the distribution cables and by correcting the power factor.

Transformers. Distribution transformers normally operate above 95% efficiency, unless they are old or are operating at very light load. Old, inefficient transformers should be replaced by new models that are more efficient. It is more efficient to run only one transformer at full load than to run two transformers at light load.

Cable Sizing. The currents supplied to the motors in any given installation will produce losses (of the I^2R type) in the distribution cables and transformers of the consumer. Correct sizing of the cables will not only allow a cost-effective minimisation of those losses, but also helps to decrease the voltage drop between the transformer and the motor. The use of the standard national codes for sizing conductors leads to cable sizes that prevent overheating and allow adequate starting current to the motors, but can be far from being an energy efficient design.

In general, in new installations it is cost-effective to install a larger cable than that required by code, if the larger cable can be installed without increasing the size of the conduit, the motors operate at or near full load and the system operate a large number of hours per year.

Power Factor Compensation. A poor power factor means higher losses in the cables and transformers, reduced available capacity of transformers, circuit breakers, and cables, and higher voltage drops. In the case of motors, the power factor is minimum at 0% load, and it increases with the load (Figure 7).

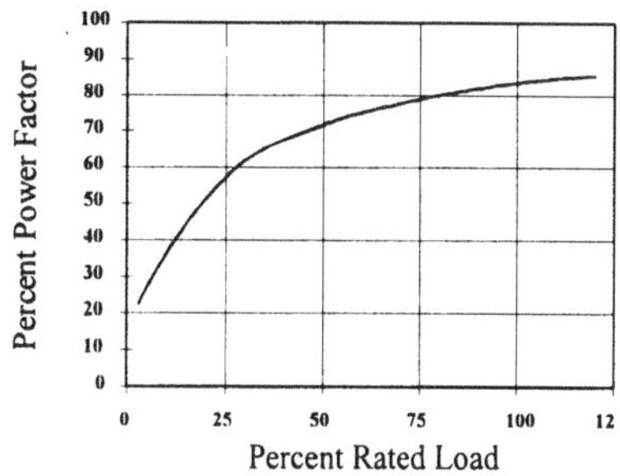

Figure 7.
Typical power factor vs load curve [4]

An oversized motor will drop significantly the power factor. Thus the proper motor size will improve the power factor. Low power factor can be corrected by using capacitors connected to the motor or at the distribution transformer and by using the correct motor size. Reactive power compensation not only reduces the losses in the network but also allows full use of the power capacity of the power system components. Additionally, the voltage fluctuations are reduced, thus helping the motor to operate closer to voltage for which it was designed.

2.4.4. Transmission System

The transmission system transfers the mechanical power from the motor to the final end-use. The choice of transmission is dependent upon many factors, namely: the desired speed ratio, motor power, layout of the shafts, type of mechanical load, etc. The most important kind of transmission types available include: direct shaft couplings, gearboxes, chains and belts.

Belts. Most motors are connected to their loads through a transmission system, very frequently through a belt. About one third of the motor transmissions in industry uses belts. Belts allow flexibility in the positioning of the motor in relation to the load. Additionally, belts can also increase or decrease the speeds using pulleys of suitable diameters. There are several types of belts namely: V-belts, cogged V-belts, synchronous belts and flat belts. While V-belts are the cheapest and are the most common type, other types can offer greater efficiency (Table 2).

TABLE 2. Comparison of belt drive characteristics [5]

	Typical Efficiency Range (%)	**Suitable for Shock Loads**	**Periodic Maint. Required**	**Change of Pulleys Required**	**Special Features**
V-belts	90-98	Yes	Yes	No	Low first cost.
Cogged-V-belts	95-98	Yes	Yes	No	Easy to retrofit. Reduced slip.
Synchro-nous Belts	97-99	No	No	Yes, with higher cost	Low-medium speed applications. No slip. Noisy.

The V-belt losses are associated with flexing 4 times per cycle, slippage and a small percentage loss due to windage. With wear V-belts stretch and need retensioning. They also smooth with wear, becoming more vulnerable to slip. Thus V-belts need regular maintenance, which is a disadvantage in relation to other non-stretch type belts. Besides, their efficiency will drop if the load is above or below the full load (Figure 8).

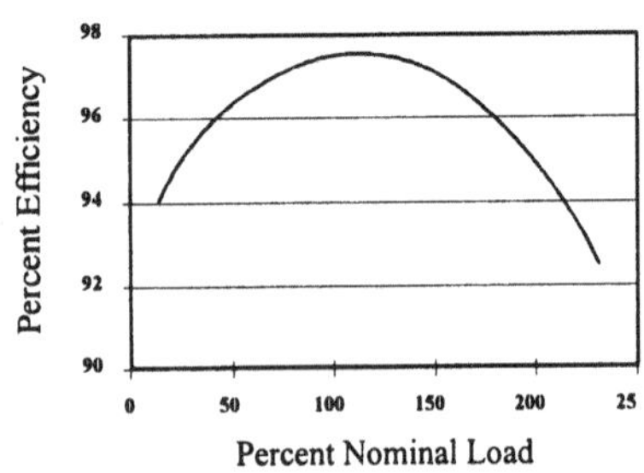

Figure 8.
Efficiency curve for a V-belt [6]

The cogged V-belts have lower flexing losses, since less stress is required to bend the belt and so they are typically 1-4 percent more efficient than standard V-belts. They can be used on the same sheaves and pulleys as standard V-belts, last twice and require less frequent adjustments. The efficiency gained with cogged V-belts is larger when small pulleys are used. Cogged V-belts cost 20-30% more than standard V-belts, but their extra cost is recovered over a few thousand operating hours.

The most efficient belt is the synchronous design, with 97-99% efficiency, because it has low flexing losses and no slippage. Synchronous belts have no slippage because they have meshing teeth on the belt and pulleys. Unlike standard V-belts that rely on friction between the belt and the pulley grooves to transmit the torque, synchronous belts are designed for minimum friction between the belt and the pulley. Due to their positive drive, these belts can be used in applications requiring accurate speed control. Synchronous belts stretch very little because of their construction, do not require periodic retensioning and they typically last 4 times longer than standard V-belts. Retrofitting synchronous belts requires installing sprocket pulleys that cost several times the price of the belt. In cases where pulley replacement is not practical or cost effective, cogged V-belts should be considered.

Gears. The selection of efficient gear drives can be a potential for important energy savings. The ratings for gear drives depend on the gear ratio (the ratio of the input shaft speed to the speed of the output shaft) and on the torque required to drive the load.

Several types of gears can be used in motor transmissions, namely: helical, spur, bevel and worm. Helical and bevel gears are the most widely used and their efficiency can reach 98% per stage (each step of reduction or increase in the shaft speed). Spur gears are used for the same purpose as helical gears but are less efficient, so they should not be used in new applications. Worm gears allow a large reduction ratio (5:1-70:1) to be achieved in a single stage. Their efficiency ranges from 55% to 94% and drops quickly as the reduction ratio increases. Thus, worm gears should be whenever possible replaced with more efficient gears as helical gears.

Chains. Unlike belts, chains typically have been used in low speed and high-torque applications. Like synchronous belts chains do not slip. A well-maintained chain may have an efficiency of about 98%, but wear can decrease this efficiency of a few points. With the exception of silent chains, chains are noisy. Chains need readjustments and adequate lubrification, which may not be easy to provide. Thus the use of synchronous belts may seem as an attractive alternative to the use of chains.

2.4.5. Maintenance Practices

Regular maintenance (such as inspection, cleaning, lubrication, tool sharpening) is essential to maintain peak performance of the mechanical parts and to extend their operating lifetime.

Lubrication. A regular maintenance with the right frequency is necessary, to reduce to the minimum the friction of the bearings. Bearing friction wastes energy, increases the motor running temperature, decreases both the motor and lubricant lifetimes. Both under or over lubrification can cause higher friction losses and shorten the bearings lifetime. Additionally, overgreasing can cause the accumulation of grease and dirt on the motor windings, leading to overheating and premature failure. The use of synthetic lubricants can achieve substantial reduction in the friction losses.

Periodic Checks. The temperature, and the electrical and mechanical conditions of a motor should be checked periodically. Additionally, the mechanical efficiency of the end-use tool (pump, fan, weaving machine, etc.) directly affects the overall system efficiency. Monitoring wear and erosion in the end-use tool is especially important as its efficiency can be dramatically affected. For example, the erosion of the pump impeller will cause the pump efficiency to drop sharply.

In general facilities with good maintenance programs will inspect the motor driven system every six months.

Cleaning and Ambient Conditions. Cleaning the motor casing, which is frequently required in some dusty industries, is also relevant because its operating temperature increases as dust and dirt accumulates on the case. The same can be said about providing a cool environment to the motor. The temperature increase leads to an increase of the windings resistivity and therefore to larger losses. An increase of 25 ^{o}C in the motor temperature increases the Joule losses by 10%.

2.4.6. Load Management and Cycling

In addition to energy savings, demand reduction can also be achieved through the use of energy-efficient motor systems. Especially in the case of large investments, the economic benefits of demand reduction should also be taken into account when evaluating the cost-effectiveness of energy conservation investments. Additionally, motor cycling and scheduling can be performed, for load management purposes, to further reduce the power demand during peak periods. Typical loads which may benefit from cycling are loads with large time constants. Such loads include refrigeration equipment, air-conditioners, heat-pumps and other curtailable loads.

2.5. SUMMARY OF TECHNOLOGICAL OPPORTUNITIES

Table 3 summarises the opportunities for improve the efficiency of electric motor systems, identified by a recent ACEEE study:

TABLE 3. Summary of electric motor system efficiency opportunities [7]

Measure	Saving Potential as % of Motor End-use Fraction	Assumptions
Energy-Efficient Motors	1-9 % for applicable motors	96 % of motor load has energy-efficient replace-ments available
Correction of motor oversizing	6-9 % for oversized motors	one-third of motor load operates a less than 50% load
Improved Drivetrains, Lubrication and Maintenance	3-7-% for all motor load	
Improved Electric Supply	1-5 % for all motor load	
Adjustable Speed Drives	15-40 % for applicable motors	25-75 % of pump, fans, compressor, and conveyor loads are applicable sites
Pump and Fan Efficiency	2-10 % for end-use motor load	
Reduced Compressed Air Leaks	50-75 % reduction in leaks	compressed air leaks are 15 % of compressor load
Compressor Controls	3-7 % for compressor load	
Air Compressor Efficiency	4-20 % for applicable load	50 % of compressor load is applicable for equipment efficiency improvement
Refrigeration Equipment Efficiency	10 % for refrigeration load	

3. Main Barriers to the Penetration of Energy Efficient Motor Technologies

Despite all the advantages that comes with the use of Energy Efficient Motor Technologies, they still have not been widely adopted. There are some barriers that can hinder the adoption of such technologies. The most important barriers to the penetration of energy efficient technologies are described below:

3.1. LACK OF INFORMATION

3.1.1. Ignorant Consumer

The implementation of energy efficiency improvements in every sector requires the awareness, support and participation of the ultimate consumer of energy. This involvement depends upon the consumer knowing about the energy technology, being aware of the possible efficiency improvements and understanding the costs and benefits of the different options. However, many consumers are quite ignorant of the possibilities of efficiency measures. Consumers should also be aware of the function of the driven system, about the range of solutions available and about the optimal design of the whole drive system adapted to the specific application.

3.1.2. Savings Difficult to Measure

It is often difficult for utility customers to know the contributions of individual end-users from the more general information typically offered on an electric bill. Utilities need to provide considerably more information and technical support if consumers are to have the necessary information to make "rational" decisions on energy use [7].

3.2. PAY BACK GAP

The "pay back gap" is a key impediment to investment in more efficient equipment. In general commercial and industrial consumers will invest in energy-efficiency improvements only if the resulting energy savings will yield a simple pay back period of two-three years. This contrasts sharply with utility investment perspectives: they invest in power plants that take up to 20 years to pay back. This gap in pay back requirements causes society to invest too little in energy savings and too much in the supply-side.

3.2.1. Capital Market Imperfections

The market for energy efficiency, determined by the decisions of millions of individual investors, operates at discount rates much higher than electric utilities cost of capital, they are able to borrow money with low interest rates (typically 4-10%). Indeed, in most cases the decision on efficiency choice is made with a discount rate several times greater than the utility cost of capital. Considering that energy efficiency can successfully replace new energy supplies, these differences in the decision criteria for investment in efficiency and in new supply means that capital will be preferentially allocated to supply.

Like an energy utility, the financial institutions that provide the capital are also supply biased. According to the conventional approach to energy followed by the financial institutions, the purpose of the energy system is to increase energy consumption which means increasing the energy supply. Efficiency improvements, which serve to reduce the supply, have been automatically ignored. Thus new power plants are built instead of investing in efficiency.

3.3. CONSUMERS BEHAVIOUR AND CONCERNS

3.3.1. Scepticism

Consumers often do not believe the technology performance claimed by manufacturers. They often claim savings which may be exaggerated under certain conditions. This creates some scepticism about technology performance.

3.3.2. Customer Indifference

There are some energy informed consumers who are indifferent to energy efficiency improvements. There are two major factors account for this attitude:

- energy cost is just one (and often not a very important factor) of many factors (such as brand names, aesthetic appeal, etc.) concerning energy equipment purchases.

- energy costs of these consumers are not so significant to motivate them to implement energy efficiency improvements, even thought the savings coming from these improvements are very important to society at large.

The following illustrative example could be useful in order to underline the importance of life-cycle costing and the tendency of users in underestimating all costs other than the direct initial investment: " After 10 years of operation at 2000 h/year, a 1,1 kW (11 kW) motor accumulates electricity costs more than 7 (respectively 14) times higher than the initial investment costs."

In developing countries, in particular, the unrealistic pricing of energy carriers serves to exacerbate this problem. Many customers preferred to pay the low initial costs of standard motors rather than the benefits of improved efficiency.

3.3.3. Risk/Corporate Image

Another important customer requirement concerns the reliability. When making a significant purchase, consumers, builders and installers all want to know that the purchase will last for at least 15 years without any problem. But some energy efficient technologies do not have such extensive track record because they are so new.

For the manufacturing process one primary and very important issue is the effect that a new and unproved technology may have in the quality of the final product. This issue is more and more important as quality is a key factor to ensure competitiveness. If it is true that the large customers are more and more receptive to the efficient technologies, their priority is still the investment in well known and reliable conventional technologies.

3.4. MOTOR MARKET CHARACTERISATION

3.4.1. Market Structure

One of the difficulties to promote EEMs is the way how motors are sold. There are three different ways to sell motors:

- through original equipment manufacturers (OEMs)
- through motor distributors or vendors
- through direct sales

Because Original Equipment Manufacturers (OEMs) compete largely on the basis of price, they tend to avoid the use of more expensive, high-efficiency motors. And because they are several steps away from the end-users in the market chain, they rarely receive direct requests for efficiency improvements.

The motor market structure is as follows:

- about 60% of all electric motors are purchased directly by OEMs, and only 40% by end users.
- about 60% of the end-user motor sales are to replace old equipment and 40% of the end user motor sales are to new installations or renovations. Maintenance staff usually make purchase decisions on replacement and they typically do not have the necessary expertise or incentives to make energy-efficient choices. Engineering staff typically make purchase decisions on new applications and because of their expertise, they are more receptive to the application of energy-efficient motor technologies.

3.4.2. Turnover Rate

The turnover rate for a given product helps determine the annual market for that product. For example: in the case of motors, which will typically last for 15 years, this means that the maximum potential market penetration for a EEM is about 6% in any year. Due to other factors, such as the rewinding, it is virtually impossible for EEMs to achieve the entire 6% penetration.

3.4.3. Distribution system

One factor that influences in a significant way a product market penetration is the capability of the product distribution system to make the product available and to supply it to the customer who wants it. If supply cannot keep up with demand, there's a potential for loosing consumer interest to the inefficient options.

3.4.4. Technical Assistance

When customers makes a purchase, they always choose a model (or brand) that ensures them a reliable repair service in the case of a failure in their new technology.

3.5. SPLIT INCENTIVES

OEMs have no motivation to buy more EEMs because they will not pay the energy bill. Their main concern is price and time of supply. Unlike OEMs, end users have a strong interest in the efficiency of the motor because they have to pay the energy bills. Because of their relative importance, the main challenge is the incorporation of EEMs by OEMs.

3.6. PURCHASE DECISION CRITERIA

Those in charge of purchasing drive power equipment are typically untrained and uninvolved in the energy-use consequences of their choices. They are quite ignorant about energy efficiency costs and benefits. They select products primarily on the basis of price (first cost) and delivery terms, leading to poor or inefficient choices.

Usually technical information related to energy efficiency is unclear and not "transparent". To understand that information requires specific knowledge from consumers that normally they do not have. As a result they often make sub-optimal choices among the set of available options.

3.7. LIMITED ACCESS TO CAPITAL

Even an informed consumer who wishes to invest in more efficient equipment may have limited access to capital, for which energy savings must compete with many other potential investments.

The typical end-user is more restrictive with capital than with operating funds. Capital expenses are generally closely scrutinised and require approvals at multiple levels in a company. To minimise capital outlay, companies tend to choose the least expensive equipment that will do the job satisfactorily. On the other hand, operating funds are easier to obtain since they are required for production. It is more easy for production staff to pay higher electricity bills every year than to get authorisation and budget for efficiency improvements.

3.8. LACK OF INTERNAL INCENTIVES

Many firms are organised in some individual departments each one having their own budget. This situation imply the partitioning costs of equipment trough the different departments: purchase department support the capital for efficiency improvements and maintenance department support the operating costs. Thus the purchase department has no incentive to invest in energy efficiency improvements that will reduce operating costs, because this practice just will benefit other department. Besides, there is a lack of internal rewards for departments and/or people to pursue energy efficient investments.

3.9. AVERSION TO DOWNTIME AND INNOVATION

Motors are routinely rewound, which can involve an outage of a day or more. In industry downtime can mean loosing thousands of dollars per hour in forgone production. Such penalties may induce an understandable aversion to downtime and cause many facility managers to shy away from new, unfamiliar technology they fear might to be less reliable than the equipment they are used to. If a EEMs is not stocked in order to save time, the user will likely buy a standard substitute or simply will repair the old motor. A common consequence of aversion to downtime is also that motors tend to be oversized for (unjustified) security reasons and therefore may run inefficiently at part-load.

3.10. REWINDING

A motor is usually repaired one or more times to extend his useful life, because it is economic in terms of first costs. Motor repair consists of rewinding (process of stripping out old windings and replacing them with new wire) and replacing bearings. In theory, rewinding can produce a motor with the same efficiency rating as it had when it was new. In practice motor efficiency is often degraded through normal rewind practices, making the initial low cost a potentially poor investment.

This practice establishes a common barrier that is difficult to overcome and became more important because there are not enough data available to precisely estimate the decline in efficiency caused by poor rewind. The experience in USA shows that on average each time a motor is rewound, its efficiency drops by 1%.

3.10.1. Repair Shops Compete on Speed and Price

When a motor fail it must be repaired as fast as possible to reduce at minimum the downtime. This situation pressures repair shops to be faster than the competiors. So repair shops must compete through price and speed. One of the most effective ways to hasten a motor rewind is to increase the temperature of the oven used to burn out the motor windings. The higher the temperature is, the more the motor core is damaged, which may lead to significant losses of efficiency and to shorten the motor life.

4. Strategies to Promote Energy Efficient Motor Technologies

Besides the actual good performance of energy efficient technologies, they have not yet been widely adopted. However, estimates proves that there is a significant potential for energy savings if efficient technologies are used in the industrial and commercial sectors by 2010. Thus can explain the actual need to promote energy-efficient motor technologies. Some strategies which can be used to overcome the identified barriers and contribute to make efficient purchase decisions are examined bellow.

The most effective way to promote energy-efficient motor technologies is through programmes that involve all the market players: utilities, trade allies, end-users, distributors, OEMs, etc. Financial incentives and coupled with a suitable regulatory framework are a more effective way to promote such technologies than individual programmes.

4.1. EDUCATION AND TRAINING

More education on energy efficiency is needed on many fronts. Motor system efficiency should be incorporated into engineering and technicians curricula. Junior engineers need one-on-one field training with experienced engineers and practising professionals should have ready access to continuing education programmes. Training of consumers is a powerful way to raise their awareness with regard to the benefits of EEMs. Training vendors so that they promote EEMs by their own desire and encourage end users to trust in vendors' advices also yield market transformation.

A good example of education and training is the documentation on the Swiss RAVEL project (technical report, exhaustive collection of demonstrated examples etc.) or the Best Practice Programme in the UK.

Most end-users and many consulting engineers lack adequate information on efficiency motors performance (and on drivepower system), economics and applications. This information gap should be filled through several means:

4.1.1. Publications

Comprehensive publications relating to motor system efficiency, (including leaflets, brochures, handbooks and videos), should be issued and distributed by manufacturers, trade associations, electric utilities, government agencies, etc.

4.1.2. Databases

With the widespread use of personal computers, computerised databases represent another type of very important source of data to make optimised choices. Databases can be available as a diskette or CD-ROM. With the proliferation of information technologies (such as Internet), another potential to promote EEMs emerge. Actions that will help to provide greater access to quality technical marketing and financial information are needed.

4.1.3. Seminars and Courses

Seminars that target engineering and maintenance staff are very important to promote EEMs. But is also important that seminars will target managers too, because management must be committed if efficiency improvements are to be implemented. If there is some kind of inducement, technical and maintenance staff is likely attend seminars.

4.1.4. Energy Audits

Detailed engineering analyses can be provided by several industrial audit programmes operated by utilities, regional energy offices, universities and private motor distributors. This procedure could increase awareness of electricity consumption and foster future measures to improve the efficiency situation.

4.2. PROVIDE TECHNICAL ASSISTANCE

Technical assistance programmes fall into two categories: calculation aids, and on-site assessment and assistance.

4.2.1. Calculation Aids

Motor systems are not properly optimised because those who make purchase decisions do not have the time or skills to do the necessary calculations. Calculation aids coupled with the motor databases address this problem with the use of reference tables, slide rules and computer programmes. This aids to estimate costs and savings for a specific application based on limited amount of information gathered by the user, and to evaluate the economics of rewind versus replacement decisions.

Programmes to assist motor maintenance practices and improve efficiency are also very important. Information on recurring maintenance tasks can be input into the computer, and work orders will be prepared by the computer according to a schedule set by the user.

Software packages covering various facets of motor systems technologies need to be available and to be distributed as well as it is important to train all the market players to use them.

4.2.2. On-site Assessment and Assistance

The simplest of these programmes feature computerised audits furnished by utilities or government energy agencies. These audits often assess opportunities for using efficient technologies. Technical assistance services that offer information and advice on efficiency improvements, without providing a full engineering audit are also a very important step.

4.2.3. Demonstration Projects

One of the key factor that limit the adoption of EEMs and energy-efficient motor technology, is the strong scepticism about their viability. People doubt about their durability, performance and real savings. To overcome these difficulties, demonstration projects should be implemented, that effectively address market deficiencies, and generate technical information about the cost-effectiveness and performance of these technologies. Demonstration provides a forum between promoters and end-users that can give some feedback to promoters about end-users needs and wants.

4.2.4. Procurement

This strategy aims to promote the appearance in the market of very high efficiency equipment with reasonable costs. A competition is promoted among manufacturers to produce the best equipment. The winning participant receives either a guarantee for a large purchase or is given a large prize ("Golden Carrot Incentives").

Procurement has successfully been applied in Europe and USA, for example to promote the introduction in the market of high efficiency refrigerators. In the case of motors procurement may be justified in the low horsepower range where substantial improvements can be achieved through the application of advanced technologies.

4.3. FINANCIAL INCENTIVES

Customers will only implement measures with two-three years pay back or less. Financial incentives are elaborated to overcome financial barriers (namely the pay back gap and the capital availability gap) to EEMs adoption and may represent one of the most cost effective way to promote energy efficient motors.

The most common financial incentives are described below:

4.3.1. Motor Rebates

Motor rebate programmes can be offered by utilities because these programmes are likely to be cost effective in most potential applications of EEMs. Rebates, in the USA, are typically about $10 per hp [3]. Programme eligibility levels and rebate formulas vary with motor horsepower and sometimes with motor speed and enclosure type. Most programmes target new applications or situations where a new motor will replace a failed motor.

4.3.2. Generic Rebates

Generic rebates reward customers for savings achieved through measures of the customers' choosing. They generally take one of the following forms:

- payments per kW or kWh saved (either in the first year or over the measure life)
- payments as a percent of measure cost
- payments to bring the simple payback for a measure down to a specific level

4.3.3. Direct Incentives to Motor Distributors

Motor distributors are well placed to understand consumer needs and wants. Perhaps the best way to involve distributors in utility programmes is to give them financial incentives (e.g. BC. Hydro's 1992 programme for EEMs offered rebates of US $293/kW saved to customers and $59/kW saved to distributors [8]).

4.3.4. Loans
Loans can compensate the limited access to capital that can prevent many customers from investing in efficiency. Subsidised loans with zero or low interest rates are particularly attractive to customers.

4.3.5. Leasing
Leasing specific pieces of energy-saving equipment also obviates searching for investment capital.

4.3.6. New Installation Incentives
New installation is a golden opportunity to use efficient technologies. To incorporate efficiency measures when a facility is built is much less expensive than retrofitting them later because extra capital, design, and installation costs are much lower for new installation.

4.3.7. Bidding
Under bidding programmes, utilities request proposals from outside parties to supply demand-side and supply-side resources. Bids are selected on the basis of price, and other factors. One purpose of bidding programmes is to allow the marketplace to determine the price of savings. Bidding programmes must address the following issues:

- the resources to be solicited and sectors to be targeted
- bid evaluation criteria
- bid ceiling prices
- mechanisms for measuring the savings.

Bidding is likely to contribute most significantly to savings in the large commercial and industrial sectors, where motors are heavily used.

4.4. LABELLING

There is a need to promote the development of uniform world-wide testing standards to measure the efficiency of motors, since testing standards are significantly different depending on the test method. Table 4 shows the efficiency determined by different methods in Japan, Europe and USA.

TABLE 4. Testing standards in Japan, Europe and USA
(Source: R. E. Osterlei, Proceedings of the National Conference on Power Transmission, Gould Inc., St. Louis, MO, 1980)

hp	JEC 37, Circle Diagram	IEC 34-2 Loss Summation	IEEE 112, Method B
5	88,8	88,3	86,2
10	89,7	89,2	86,9
20	91,9	91,4	90,4
75	93,1	92,7	90,0

Labelling helps to bridge the information of consumers about motor performance. Labelling also increases competition among manufacturers to raise the efficiency over the whole motor range.

Motor testing and labelling can help consumers make informed choices about motor efficiency. Labelling also should help consumers to compare relative performance against models from other manufacturers.

The label also should inform about:

- equipment efficiency (Ideally at 25%, 50%, 75%, 100% load and no-load losses)
- typical operative costs with a range of electricity prices

If available (in Europe it is not the case), motor efficiency labelling is normally only given for full-load operation. As a motor seldom operates at full-load all the time, it would be desirable to have motor efficiency labelling covering also partial load conditions (no-load, 25%, 50% and 75% load). This information would help to optimise motor selection as a function of the load diagram of the motors. The no-load losses would help to verify if the rewinding work in a failed motor is carried out in a proper way.

Motor labels will likely influence the purchase decision maker. Labels can be useful in the used motor market by telling customers the motor's nameplate efficiency when it was new. On the other hand, efficiency information about EEMs, on catalogue listings can be very useful in the new motor market.

4.5. VOLUNTARY AGREEMENTS

This strategy involves agreements between manufacturers, trade associations and regulatory authorities, to establish minimum efficiency energy levels that meet the desired performance level of contenders. On the end-use side, a similar effect can be reached by Bench-Marking-Systems of Associations. These systems allow comparison of the relative ranking in regards to energy/electricity use in comparison to other companies etc. The results of energy audits (mentioned in section 4.1) can be used for this purpose.

One successful example of voluntary agreements is the Energy Star Programme applied to personal computers, monitors and printers.

4.6. EFFICIENCY STANDARD

An important market transformation tool arrives through the establishment of minimum efficiency standards for motors. Standards define minimum efficiency levels and help to overcome the following barriers:

- lack of information of energy performance
- market inertia to introduce high efficiency equipment
- split incentives barrier
- energy costs have little influence on the buyer's decision
- remove inefficient equipment from market

This measure has been already adopted with success in the British Columbia - Canada in January 1995, and it was extended to other Canadian provinces in January 1996. In USA it will be enforced in October 1997, through the Energy Policy Act-EPAct, mandated in October 1992. Table 5, shows the minimum nominal full-load efficiency of electric motors* mandate by EPAct:

TABLE 5. Minimum nominal full-load efficiency of electric motors
* - 4-Pole closed motors, TEFC

Motor Horsepower	Efficiency (%)
1	82.5
5	87.5
10	89.5
25	92.4
50	93.0
100	94.5
200	95.0

These standards will particularly be effective for OEMs, obliging them to incorporate EEMs in their equipment. These standards recommend or require motors to exceed specific efficiency levels, which vary with motor size and with motor speed and enclosure type.

4.7. UTILITY INTEGRATED RESOURCE PLANNING (IRP)

The increasing concern about the environment consequences of electricity production and transportation (e.g.: global warming and acid rain), and the frequent strong local resistance to new power plant construction has lead utilities to work closely in cooperation with customers: utilities can cooperate to reduce the customers' electrical demand for energy and power. One of largest potential for electricity savings is associated with drivepower.

Therefore, utility promotion of energy efficiency among their customers through IRP programme can be one important cost-effective way to promote EEMs.

4.8. BEST REWIND PRACTICE

It is widely accepted that typical rewind practices cause damages in the motor and reduce its efficiency. Because of that, some utility programmes encourage consumers to substitute a failed motor with an energy efficient motor. But on the other hand this perspective is in conflict with the interests of motor repair/rewind shops. The solution is to develop best practices, perform training of staff and encourage motor repair shops to provide quality rewind services.

A certification programme for rewind shops is another option worth exploring: to be certified, shops would have to use approved procedures.

4.9. SUBSTITUTE STANDARD MOTORS BEFORE THEY BREAK

Lots of standards old motors are still operating world wide. If economical perspective is considered, it will be preferable to substitute some of this old motors by new EEMs. Actions are needed to encourage motor users to replace old inefficient motors by EEMs. For example the assessment of existing industrial systems may show that current processes requirements can be more efficient and cost-effective by the replacement of existing equipment [9].

4.10. CASE-STUDY OF B.C. HYDRO

All the identified strategies can produce the motor market transformation, but while enforced regulation and financial incentives will guarantee a large scale penetration of efficient technologies, educational programmes alone will produce marginal changes.

A combination of strategies may be used to transform the motor market. An outstanding example of motor market transformation has occurred in Canada in the service territory of British Columbia Hydro (Figure 9).

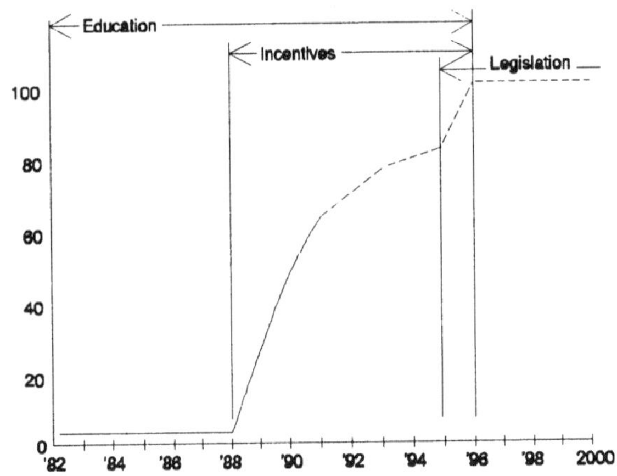

Figure 9.
Example of successful market transformation [8]:

B.C. Hydro's programme of education and financial incentive increased the market share of efficiency motors from 3% to 60% in a period of three years. The utility is now supporting legislated efficiency standards to sustain the market, and will ramp down the programme as market forces and standards take effect.

5. Conclusions

This paper has shown that there are significant opportunities to improve the efficiency of the motor system.

Speed controls, specially Variable Speed Drives, and Energy Efficient Motors are by far the most effective way to produce energy savings. The development of advanced motors and drives in order to reduce the cost of inverter driven induction motors is an important step to improve the penetration of energy efficient motor technologies.

High quality power supply, careful attention to harmonics, proper equipment sizing, attention to power factor and distribution losses, optimised transmission systems, careful maintenance of the entire drive power system and good load management practice are also important factors that influence the reliability, the longevity and the efficiency of a motor system.

However it is important to note that despite efficient technologies being already available in the market they are not widely used in most EU countries, because there are some important barriers to their large-scale penetration. Lack of information, the pay back gap, consumers behaviour and concerns, the motor market structure, purchase decision criteria, limited access to capital, lack of internal incentives, aversion to downtime and innovation, and rewinding criteria were identified as the main barriers.

Some possible strategies that can be used to overcome the barriers were also identified. Possible strategies include: education and training, technical assistance, financial incentives, labelling, voluntary agreements, efficiency standards, and demand-side management options. While enforced regulation and financial incentives (e.g.: tax incentives for those companies who buy energy-efficient motors) will guarantee a large scale penetration of EEMs, educational programmes alone will produce marginal changes. Experience in North America shows that a combination of strategies is the most effective way to transform the market towards energy efficiency technologies.

6. Bibliography

[1] De Almeida, A. T., New Developments in Energy-Efficient Electric Motor Drive Systems, Department of Electrical Engineering, University of Coimbra, Coimbra.

[2] Babb, Michael, Premium Efficiency Motors Promise to Save Billions, *Control Engineering*, May, 1995

[3] Nadel, S., Shepard, M., Greenberg, S., Katz, G., de Almeida, A., (1991), *Energy-Efficient Motor Systems*, American Council for an Energy-Efficient Economy, Washington, D.C., 1991.

[4] De Almeida, Aníbal, and Steve Greenberg, Technology Assessment. Energy-Efficient Belt Transmissions, Berkeley, Calif., Lawrence Berkeley Laboratory, 1992

[5] Almeida, Aníbal T., Greenberg, Steve, (1995) Energy-Efficient Belt Transmissions, *Energy and Buildings*, March, 1995

[6] Elliott, R. N., (1995), Energy Efficiency in Electric Motor Systems, American Council for an Energy-Efficient Economy, Washington, D.C., April, 1995.

[7] C. Hydro, (1992), Power Smart High-Efficiency Motors, Profile #38, 1992.

[8] Motor Challenge, (1995), Roundtable on Market Transformation Strategies for Industrial Electric Motor Systems - Record of Meeting, U. S. Dep. of Energy Motor Challenge, Chicago, April, 1995.

[9] Pilot-project Least Cost Planning, SAVE DGXVII-Commission of the European Community, 1994

[10] Elliott, R. Neal, Electricity Consumption and the Potential for Electric Energy Savings in the Manufacturing Sector, April, 1994

[11] DSM Pocket Guidebook - Commercial technologies, Solar Energy Research Institute for Western Area Power Administration and U.S. Department of Energy, 1991

[12] Miller, Peter M., Eto, Joseph H., Geller, Howard S., The Potential for Electricity Conservation in New York State - Final Report, American Council for an Energy-Efficient Economy (ACE3) for Niagara Mohawk Power Corporation and New York State Energy Office, September, 1989

[13] Lamarre, L., (1994), Efficiency a Hard Sell, *EPRI Journal*, October/November, 1994.

[14] Cheek, K.F., et al., (1995), Impact of Motor Failures on Payback Periods, In Proceedings of the Industrial Energy Technology Conference, Houston, April, 1995.

[15] Motor Challenge, (1995), Roundtable on Market Transformation Strategies for Industrial Electric Motor Systems - Background, U. S. Dep. of Energy Motor Challenge, Chicago, April, 1995.

[16] Baldwin, S. F., (1989), Energy-Efficient Electric Motor Drive Systems, Johansson, T. B., Bodlund, B. and Williams R. H. (eds.), *Electricity*, Lund University Press, Sweden, pp. 21-58.

[17] Levine, M., *et al*, (1994), Energy efficiency, Market Failures, and Government Policy, *Lawrence Berkeley Laboratory Report No LBL-35376 UC 350*, Berkeley, March, 1994.

[18] Reddy, A. K. N., (1991), Barriers to Improvements in Energy Efficiency, Butterworth-Heinemann, *Energy Policy*, December, 1991.

[19] Wiel, H. S., (1991), The Electric Utility as Investment Bank for Energy Efficiency, *The Electricity Journal*, Mai, 1991.

[20] Macro Consulting, Inc., (1994), Overcoming Barriers to the Diffusion of Efficient Technologies, *EPRI Technical Report No TR-103527,* March, 1994.

[21] Reddy, A.K.N., (1991), Barriers to Improvements in Energy Efficiency, *Lawrence Berkeley Laboratory Report No LBL-31439, Berkeley*, October, 1991.

[22] Totten, M., Settina, N., (1993), Energy Wise Options for State and Local Governments 2ª ed., Solstice, Center for Policy Alternatives, January, 1993.

LOW ELECTRICITY HOUSES - A REAL CASE: THE PASSIVE HOUSE IN DARMSTAD-KRANICHSTEIN

WOLFGANG FEIST
Passivhaus Institut
Steubenplatz 12
D - 64293 Darmstadt, Germany

WITTA EBEL
Institut Wohnen und Umwelt
Annastrasse 15
D - 64285 Darmstadt, Germany

1. Introduction and Summary

The goal of the *"Electricity Saving in the Passive House"* demonstration project was to prove that it is possible, using technology available today, to reduce the electricity consumption for end-uses in the residential sector to one third of the value of comparable households, with the associated growth in gas consumption for heat end-uses amounting to less than 15 %. The expectations were confirmed by the measured results:

Compared to a reference household in the building stock, the model household needs

- 65 % less electricity
- 61 % less primary energy and gives rise to
- 61 % less CO_2 emissions.

The savings were achieved through extra investments in electricity-saving systems alone.

The systems were tested under normal conditions of use.

The results of the demonstration project are universally utilizable: The equipment of the demonstration household shows what the final state with regard to efficient use of electricity can look like in *all households* after the complete replacement of all the less efficient appliances currently still in use, if when each appliance needs to be replaced the most efficient model on the market is purchased. As a rule, such replacement purchases are cost-effective even without initial financial support.

S. Nørgård et al. (eds.), Development with Sustainable Use of Electricity, 187-204.
Kluwer Academic Publishers.

2. Reference values: Household electricity consumption in building stock

2.1. INDEXES

The states of former West Germany serve as the reference standard for the built passive house in Darmstadt-Kranichstein. The consumption data can be put in relation to households, persons or floor space. The sources of the data are: [1],[2],[3] and updates. In the year 1990, there were in the states of former West Germany 26. 1 million occupied dwellings (excluding hostels/homes), with a total living floor area of 2257 million i& [4].

Table 1. Basic data

states of former West Germany	1991	1992	1993	1994
Inhabitants (millions)	64.07	64.87	65.53	65.90
Households (millions)	28.6	29.0	29.5	29.9
Living floor area (million m^2)	2275	2296	2323	2357
Household electricity consumption (TWh/a)	104.29	104.79	107.70	107.41
of this, for electric thermal storage (TWh/a)	20.70	19.80	20.20	19.00

Table 2. Electricity consumption indexes

states of former West Germany		1991	1992	1993	1994	Average
Electricity consumption per person (kWh/(Pa))		1628	1615	1643	1630	1629
Electricity consumption per household (kWh/a)		3648	3608	3651	3592	3625
Electricity consumption indexes:	total household electricity consumption	45.8	45.6	46.4	45.6	45.9
Consumption over floor space (kWh/(m^2a))	- without electric thermal storage	36.7	37.0	37.7	37.5	37.2
	- without electric thermal storage and water heating	30.9	31.1	31.7	31.6	31.3

Reference households suitable for comparison with the passive house are those without electric thermal storage and water heating. The electricity consumption index exclusive of these end-uses figured on average over the 1991-1994 period 31.3 kWh/(m^2 a). In the passive house, the number of persons per household is relatively high. The dependence of electricity consumption upon household size must therefore also be examined.

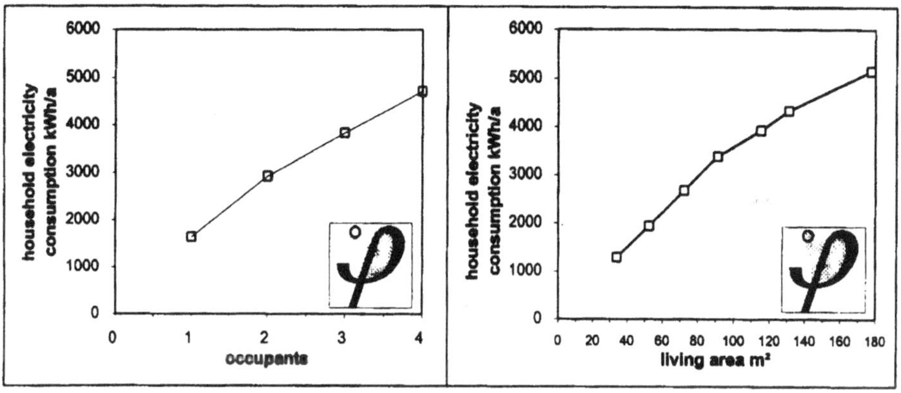

Figure 1.
Electricity consumption (without thermal storage) per household

For the size of the passive house (156 m^2 and 3.5 persons per dwelling), interpolation leads to the following reference values:

Household electricity consumption	4275 kWh/a
- per person	1220 kWh/a
- index per unit floor area	**31,1 kWh/(m^2a)**

In the following, this index is taken as the reference standard.

2.2. BREAKDOWN OF ELECTRICITY CONSUMPTION ACCORDING TO END-USES

Table 3 shows the average electricity consumption for the presently commonplace end-uses. The consumption figures refers to such households that own the respective appliances, and apply to the 1992/1993 period (source: [5], [6], [3], updates). It must be taken into consideration that the average household size in this Table is 2.23 persons.

Table 3. Specific electricity consumption and saturations of electric household end-uses

Appliance / End use	Saturation	Specific electricity consumption (kWh/a)
Electric thermal storage units	8.6%	8100
Electric water heating	36%	1300
Pump of heating system	75%	270
Domestic hot water circulation	35%	170
Refrigerator	81%	300
Combined refrigerator/freezer	24%	500
Freezer	65%	450
Dishwashing machine	39%	430
Clothes washer	100%	220
Clothes dryer	25%	440
Electric cooker	80%	380
Lighting	100%	300
Television	96%	100
Small and unspecified appliances	100%	300

3. Efficient use of electricity in the passive house

3.1. THE MODEL PROJECT: ELECTRICITY SAVING IN THE PASSIVE HOUSE

By definition, a passive house is a building in which the total end-use energy consumption index (the total of space and domestic hot water heating, and all household electricity uses) is at most in the order of 31 - 32 kWh/(m^2a) [7], [8].

As space heating requires ca. 10 kWh/(m^2a) and the supplementary heating of domestic hot water requires a further 6 kWh/(m^2a), 50% of the electricity for all other purposes had to be saved in order to achieve this index. To this end, the most energy-efficient appliances on the market were sought and purchased. In addition, gas cookers were purchased, and the clothes washers and dishwashing machines were connected to the domestic hot water system.

This demonstrates the final state with regard to efficient electricity use that is feasible in all *households* after replacement of all the less efficient appliances that are presently still in use, if at the point in time when each appliance needs to be replaced the most efficient one currently available on the market is purchased. The special feature of the demonstration project in the passive house is that the new purchase of the most efficient systems was brought ahead to the point in time of first occupation of the house.

In order to keep track of electricity consumption, the five central meters (four dwellings, and one for joint electricity uses) of the supplier and four submeters for individual gas consumption were regularly read by the occupants. These data provide the basis of the overall evaluation for the house.

In addition, the following were continuously registered within the measuring programme

- electricity consumption of the joint heating system;
- electricity, hot and cold water consumption of the two jointly used clothes washers;
- for one of the 4 dwellings: electricity consumption for refrigerator, freezer, dishwasher, computer and ventilation;
- electricity consumption for measuring system itself (is not considered in the assessment of the household's electricity consumption).

The data acquisition was performed by the ebök in Tübingen. The data were supplemented by individual measurements of the occupants.

3.2. MEASUREMENT RESULTS: ANNUAL ELECTRICITY CONSUMPTION IN THE PASSIVE HOUSE

The passive house in Darmstadt [9] is occupied by 14 persons in 4 dwellings with 156 m 2 living space each. To receive an aggregate figure, the electricity consumption levels of the four first consumption periods from October 1991 (commencement of occupation) to September 1995 were evaluated: The total electricity consumption of the 4-family house figured, including consumption for the measuring system, 8759 kWh/year. The greater part of the measuring system consumption is registered by a submeter: 1073 kWh/a. In addition, 5 data loggers were installed in the first year and 4 from the second period onwards, with a measured continuous power input of 10 W each. The consumption of these loggers was deducted from the total consumption figure. 3 further data loggers were only operated during the fourth measurement period for four months each.

Some further minor items of consumption for the measuring system (e.g. three-phase meters) were not registered separately, and are therefore contained in the following household electricity consumption figures.

The total electricity consumption then breaks down as follows:

- electricity consumption in all household incl.ventilation	6056 kWh/a
- total joint electricity consumption	1235 kWh/a
- measuring system consumption	1469 kWh/a

In the following, the measuring system consumption is not considered. We thus arrive at a total household electricity consumption of 7281 kWh/a or 1820 kWh per year and dwelling.

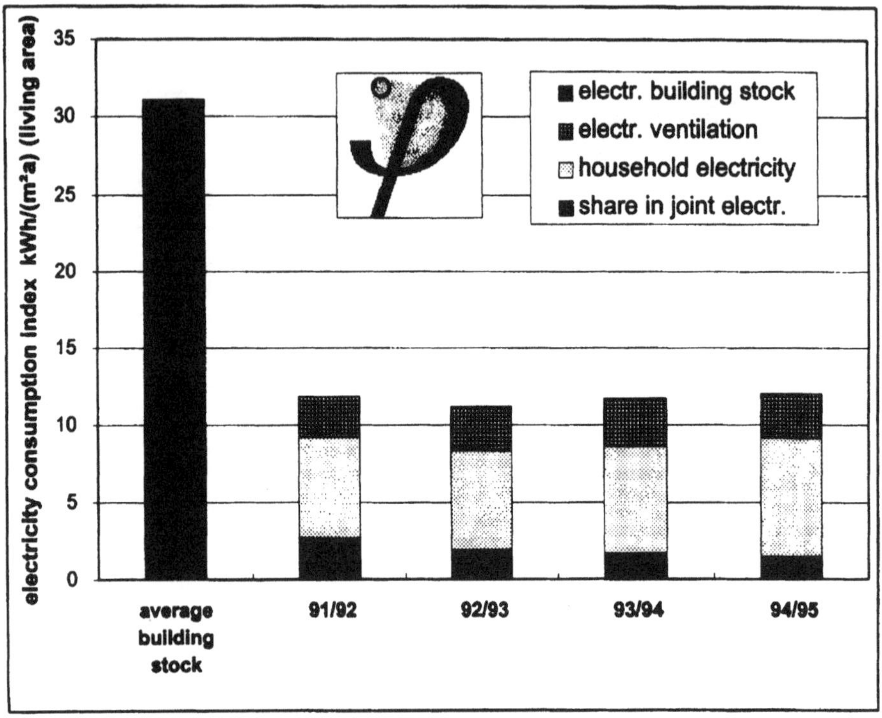

Figure 2.
Electricity consumption in the passive house

This is only half of the average household electricity consumption in the states of former West Germany, whereby it must further be considered that the average household size in western Germany is, at 2.2 persons, substantially smaller than in the passive house (3.5 persons).

Per person, the electricity consumption in the passive house figures 520 kWh/a compared to 1320 kWh/a in the building stock (without electric thermal storage).

As a function of living floor space, the average electricity consumption figures

11,65 kWh/(m²a)

With the reference value of 31.1 kWh/(m^2a) determined in Section 2 above, we find

savings: 62.4%.

The ventilation system accounted for 450 kWh or 2.9 kWh/(m^2a).

3.3. COOKING GAS CONSUMPTION

Over the 4 measurement periods, the average cooking gas consumption figured 393 kWh per dwelling and year, or 2.5 kWh per m^2 living space and year.

This natural gas consumption corresponds to the electricity consumption that would have arisen with an alternatively used electric cooker. If we add the cooking gas end-use energy to the electricity consumption (without further considering the better primary energy balance), then the energy consumption index for the total of electricity and cooking gas still only figures 14.2 kWh/(m^2a), i. e. savings of 54.3 % compared to the reference value (31.1 kWh/(m^2a).

4. The individual end-uses

The electricity consumption of individual appliances was only registered in detail for one of the four dwellings. The breakdown of electricity consumption is therefore presented for the exemplary case of this one dwelling.

4.1. REFRIGERATION AND FREEZING

As a refrigerator, an appliance supplied by the Gram company and generally available on the market was used. This LER (Low Energy Refrigerator) was developed at the Technical University in Denmark [10], [11], [12]. The stated standard consumption figures 0.24 kWh per day. The measured electricity consumption figured 101 kWh/a in the 1992/1993 measurement period. These appliances were used in all 4 of the households.

Two larger freezers were used jointly by two households each. The freezers are located in the joint cellar, but are registered by two of the private meters. The freezer measured here was purchased upon commencement of occupation, and one of the most energy efficient models on the market was selected. In the 1992/1993 measurement period, the measured electricity Consumption of the freezer figured 233 kWh. As the neighbouring household uses the freezer to the same extent, only half of this electricity consumption is assigned to the one household.

4.2. DISHWASHING

When the occupiers moved in 1991, dishwashing machines with low standard consumption levels (1.6 kWh at 65°C, 12 standard covers) were purchased. These appliances were connected to the domestic hot water system. In the 1992/1993 measurement period, an average of 0.62 kWh electricity was consumed per washing. The mainly used mode was 50°C without pre-rinsing. In this mode, the water consumption figures 18.5 l per washing. It was not possible to conduct comparative measurements with appliances connected to cold water. Instead, measurement results of Stiftung Warentest (the major German consumer goods testing institution) were used:

For the appliance used in the passive house, the "AEG Öko Favorit", the following figures were measured (reported in the 12/93 issue of the Stiftung Warentest's journal, "test"):

65°C, with pre-rinsing: 1.71 kWh electricity, 22 l water 50°C, with pre-rinsing: 1.60 kWh electricity, 23 l water

Dispensing with the first rinsing cycle reduces the water consumption by 4.5 l or ca. 20%. The electricity consumption in this mode then figures about 1.25 kWh. It follows that the connection to domestic hot water saved about half of the electricity consumption. The remaining measured electricity consumption figured 74 kWh for a total of 116 washings in 1992/1993.

4.3. VENTILATION SYSTEM

The experience made to date shows that a ventilation system is indispensable for satisfactory interior air quality. The associated additional electricity consumption must therefore in future always be taken into account.

The four dwellings in the passive house are equipped with four identical ventilation systems. The fresh air is first preheated in a subsoil heat exchanger and then flows over a counterflow heat exchanger in which it is further heated (by the spent exhaust air extracted from the bathrooms and kitchen) before being conveyed to the living rooms and bedrooms [13]. The counterflow heat exchanger available on the market was equipped with two ventilators with 100 W each. As this would have led to excessive electricity consumption, the two ventilators were replaced by direct-current (DC) blowers with electronic commutators [14].

The consumption of the two blowers inclusive of frequency conversion was registered by an additional measurement [15]. In the three selectable settings it figured:

- weak 12 W
- normal (mainly used setting) 28 W
- strong 82 W

In automatic operation, both the volume and the direction of the air flows are controlled in order to maintain a certain air quality. The ventilation control with microprocessor and the CO_2 and moisture measurement (together 19 W) and the motorized air throttle adjustment require an additional electricity consumption. The annual electricity consumption of the ventilation system figures 449 kWh/a. The heat recovery and the ventilation control system together save, however, approx. 4500 kWh/a space heating i.e. ten times the power input. This ventilation system thus provides high primary energy savings.

4.4. MISCELLANEOUS PRIVATE HOUSEHOLD ELECTRICITY CONSUMPTION

The computer in the study room was registered with a meter of its own; its contribution to consumption was negligible in the registered period.

A further meter measures the electricity that is consumed within the insulated building envelope. After deduction of the measured electricity consumption of the ventilation system, refrigerator, dishwasher and computer, 156 kWh remain for lighting and small appliances operated within the dwelling.

For the lighting, fluorescent lamps were exclusively used in the model house. The power inputs of the individual lamps figured 7 W, 9 W, 11 W and 15 W. The fluorescent lamps replace filament lamps with power inputs of 25 W, 40 W, 60 W and 75 W. From the ontimes and the known wattages, we calculate an annual consumption

- for lighting 67 kWh/a
- for the logged appliances 54 kWh/a.

It follows that 35 kWh/a were not registered within the insulated building envelope (standby consumption, forgotten lamps and appliances, errors in the extrapolation of the manually recorded period to the year, and measurement errors of the electricity meters). Beside the measured electricity consumption within the insulated building envelope and the electricity consumption of the freezer in the cellar, 107 kWh/a remain for

- lighting of cellar and glass porch
- operation of tools and appliances in the private cellar and glass porch
- measurement differences
- unregistered consumption of the measuring system.

4.5. ELECTRICITY CONSUMPTION FOR CLOTHES WASHING AND DRYING

All the households in the passive house jointly use two washing machines in the cellar, whose electricity consumption is registered by the joint electricity meter and is not assigned to any of the parties. One of the two washing machines has both a cold water and a hot water connection. The other washing machine has only one connection, but is fitted with an external control device that mixes cold and hot water for the washing cycle depending upon the temperature setting. For test purposes, the washing machine was connected in the first months (winter 1991/1992) only to cold water, i.e. the control device was not used. In January/February 1992 the occupiers logged the washing cycles that were run. The measured consumption levels were later compared with those of January/February 1993, when the control device was installed and the boundary conditions were similar.

Table 4. Mean consumption levels for washing machine 1 using control device and two connections (cold water and hot water)_compared to operation using only cold water connection

Washing machine 1		Washing cycle (temperature setting)		
Electricity consumption (kWh/washing cycle)		30°C	60°C	95°C
Cold water: electricity		0.66	1.67	2.07
Cold and hot water	electricity	0.26	1.0	1.52
	hot water (litres per cycle)	10	15	15
	electricity savings	61%	40%	27%

The consumption of the two washing machines was registered by submeters. The total consumption figured 318 kWh/a, of which 79 kWh/a were attributable to the model household. Natural gas was further consumed for heating the hot water. This is evaluated in the overall balance in Section 5 below.

4.6. MISCELLANEOUS JOINT ELECTRICITY CONSUMPTION

The consumption for the heating system (incl. hot water) was registered separately by a submeter. In the 1992/1993 measuring period this totalled 333 kWh. The circulation pump was controlled with a timer. The power input measurement and the known on-times give an annual electricity consumption of 64 kWh. 269 kWh remain for the water boiler including pump of the heating system. The electricity consumption for the solar thermal system was also measured with a submeter; in 1992/1993 this figured 105 kWh/a. 6600 kWh gas were saved with the solar system. The consumption of the rainwater pump was extrapolated from individual measurements; this figured 126 kWh/a. The remaining joint electricity consumption of 244 kWh is distributed among the lighting of the joint cellar and small appliances (bells, intercoms, unregistered three-phase meters, heat meters).

Table 5. Overview of the electricity consumption in the model house

Balance of test house	Electricity consumption 1992/1993 [kWh]		
	Consumption of test house	Share of model dwelling	Consumption of model dwelling
Electricity for water boiler incl. pump of heating system	269	0.25	67
Circulation pump	64	0.25	16
Refrigeration	101	1	101
Freezing	233	0.5	117
Dishwashing	74	1	74
Clothes washing	318	0.25	79
Lighting (private)	95	1	95
Lighting (joint)	49	0.25	12
Computer	12	1	12
Small appliances (logged operation)	54	1	54
Miscellaneous end uses (private)	114	1	114
Miscellaneous end uses (joint)	196	0.25	49
Further measuring devices, in particular meters	88	1	88
Further measuring devices (joint)	95	0.25	24
Further end uses: Rainwater pump, solar plant pump	230	0.25	58
Ventilation system	449	1	449
Total			1409

4.7. TOTAL ELECTRICITY CONSUMPTION OF THE MODEL HOUSE AND A REFERENCE HOUSEHOLD - ACHIEVED ELECTRICITY SAVINGS ACCORDING TO END-USES

The measured model household has 4 occupants, and is thus not comparable to the average set out in Section 2 above. In order to compare the total and individual appliance/system consumption levels, a conventional 4-person household is assumed that commands over the same equipment with appliances as the model house, insofar as such equipment is commonplace.Table 6 gives an overview of the specific consumption levels of a 4-person household in the building stock, the consumption levels achieved in the passive house (model dwelling), the resultant savings percentages, and, for purposes of comparison, the theoretical savings potential stated in [3]. The comparison shows that in the model dwelling the forecast savings were achieved for all end-uses.

Table 6. Savings achieved in the model dwelling

Specific consumption levels of a 4-person household kWh/a	Model dwelling	Building stock	Savings (as measured)	For comparison: Previously determined theoretical savings potential
Electricity for water boiler incl. pump	67	330	80%	80%
Circulation pump	16	210	92%	88%
Refrigeration	101	380	74%	70%
Freezing	117	510	77%	77%
Dishwashing	74	500	85%	32%
Clothes washing	79	400	80%	41%
Lighting	107	500	79%	75%
Electronics, small appliances, unspecified	341	550	38%	36%
Pumps for solar thermal and rainwater	58	0	0%	0%
Ventilation system	449	0	0%	0%
Total electricity (without cooking, electric water heating, electric thermal storage)	1409	3380	58%	55%

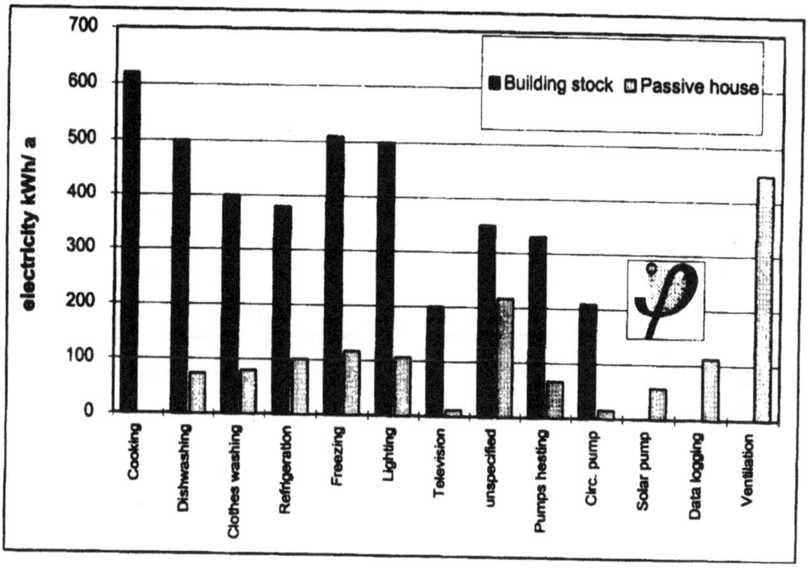

Figure 3.
Electricity consumption according to end-uses in the passive house

5. Fuel consumption for cooking, dishwashing and clothes washing

5.1. GAS USE FOR COOKING

The values measured in all 4 dwellings (averaging 393 kWh/a) suggest that the end-use energy consumption of the gas cooker tends to be lower than when cooking with electricity, due to the better controllability of the former. In the absence, however, of reliable comparative data, here only the primary energy and emissions balances shall be compared that would have arisen with the same end-use energy input:

Table 7. Primary energy and emissions balances for cooking with electricity and gas

kWh/a	Electric cooker	Gas cooker	Savings
Electricity	335		
Natural gas		335	
Primary energy	1073	362	66%
CO_2 (kg/a)	223	81	64%

5.2. HOT WATER USE FOR DISHWASHING

In addition to the electricity use (averaging 0.62 kWh), 18.5 l hot water were required per washing, and the heat input (ex circulation piping) of this figured on average 0.83 kWh per washing.

Table 8. Savings effect in dishwashing through hot water connection

Washing machine per washing	without hot water connection	with hot water connection: only gas boiler	with hot water connection: PH: gas boiler + solar thermal
Electricity consumption (kWh)	1.25	0.62	
Hot water consumption (l)	18.50	18.50	
Tapped heat quantity (from circulation piping)	0.00	0.83	
Gas consumption (kWh H_u)	0.00	1.22	0.42
Primary energy consumption (kWh)	4.00	3.31	2.43
CO_2 emissions (g)	833	709	514
Savings rate: electricity	0%	50%	50%
Savings rate: primary energy	0%	17%	39%
Savings rate: CO_2 emissions	0%	15%	38%

5.3. HOT WATER USE FOR CLOTHES WASHING

The two washing machines used by all 4 households are located in the joint cellar. The total electricity consumption of these machines figured 318 kWh. In addition, 6160 litres of hot water were used. The heat input ex circulation piping figured 277 kWh. Assuming 5 % tapping losses, the amount of electricity that would alternatively have had to be used with only cold water connection would have figured 581 kWh.

Table 9. Aggregate balance of two washing machines in the 1992/1993 measurement period

Balance of washing machines 1992/1993 (4 dwellings)		without hot water connection	with hot water connection	
			gas boiler	passive house: gas boiler and solar thermal
Electricity consump. (kWh/a)		581	318	
Hot water consump. (l/a)			6160	
Heat intake (kWh/a)			277	
Electricity+heat (kWh/a)			595	
Gas consump. (kWh/a)			407	139
Primary energy (kWh/a)		1858	1456	1167
CO_2 emissions (kg/a)		387	310	245
Savings rate	electricity	0%	45%	45%
	primary energy	0%	22%	37%
	CO_2	0%	20%	37%

6. Electricity, primary energy and emissions savings in the model house

As presented in Section 5, the natural gas use in the model house figured:

Cooking:	335 kWh (100% of the employed end-use energy)
Dishwashing:	48 kWh (40% of the employed end-use energy)
Clothes washing:	35 kWh (30% of the employed end-use energy)

In Table 10 and Figure 4, the electricity consumption levels of the model house are compared with an equally equipped 4-person household (including cooking) in the building stock. The primary energy and CO_2 balances shown in Table 11 take the additional natural gas consumption for cooking, dishwashing and clothes washing into consideration.

Table 10. Electricity savings and additional gas consumption in the passive house

Consumption in the model house	Stock	Passive house		Additional gas consumption	
kWh/a	Electricity	Electricity	Savings rate	kWh/a	
Electricity for water boiler incl. pump	330	67	80%	0	0%
Circulation pump	210	16	92%	0	0%
Refrigeration	380	101	74%	0	0%
Freezing	510	117	77%	0	0%
Cooking	620	0	100%	335	54%
Dishwashing	500	74	85%	48	10%
Clothes washing	400	79	80%	35	9%
Lighting	500	107	79%	0	0%
Electronics, small appliances, unspecified	550	341	38%	0	0%
Pumps for solar thermal system and rainwater	0	58	____	0	____
Ventilation system	0	449	____	0	____
Total	4000	1409	65%	419	10%

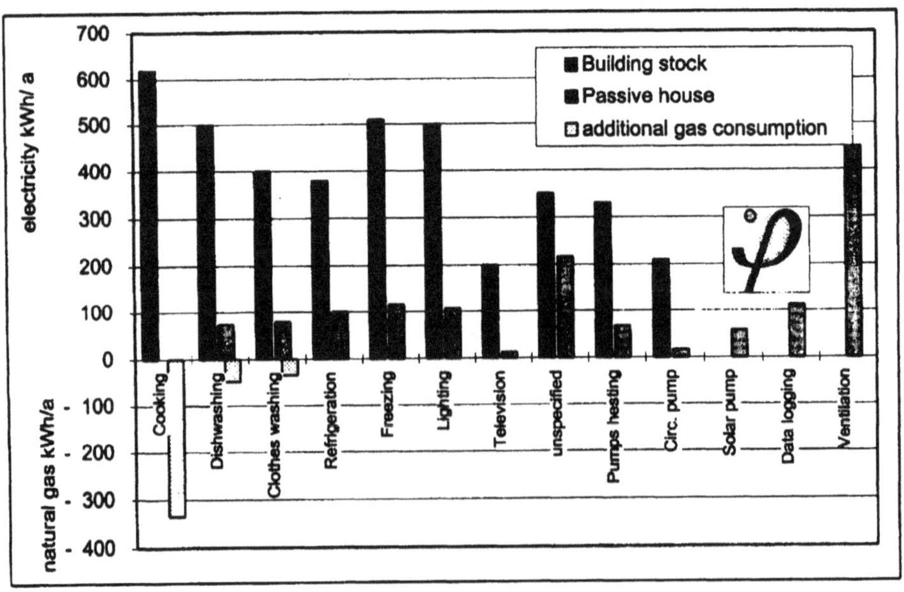

Figure 4.
Electricity and fuel consumption in the passive house

Table 11. Primary energy and CO_2 savings of electricity end-uses and cooking in the passive house

End use	Primary energy (kWh/a) Stock	Model dwelling	Savings rate	CO2 (kg/a) Stock	Model dwelling	Savings rate
Electricity for water boilers incl. pump	1056	215	80%	220	45	80%
Circulation pump	672	51	92%	140	11	92%
Refrigeration	1216	322	74%	253	67	74%
Freezing	1632	373	77%	340	78	77%
Cooking	1984	362	82%	413	81	80%
Dishwashing	1600	288	82%	333	61	82%
Clothes washing	1280	292	77%	266	61	77%
Lighting	1600	343	79%	333	71	79%
Electronics, small appliances, unspecified	1760	1093	38%	366	227	38%
Pumps for solar thermal and rainwater	0	184	___	0	38	___
Ventilation system	0	1438	___	0	299	___
Total primary energy/CO_2	12800	4962	61%	2664	1040	61%
Primary energy index	kWh/(m²a)	31.8				

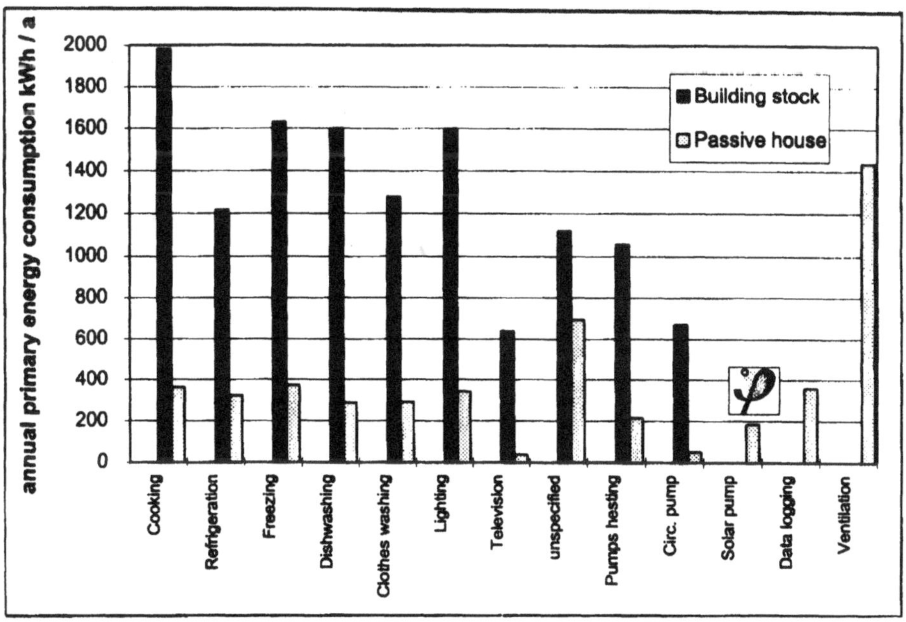

Figure 5.
savings in the passive house: comparison between model house and a correspondingly equipped 4-person household in the building stock

7. Literature:

[1] Statistisches Bundesamt: Fachserie 5, Bautäigkeitt und Wohnungen

[2] VDEW (Vereinigung Deutscher Elektrizitätswerke): (1) Die öffentliche Elektrizitätsversorgung.
(2) Ergebnisse der Haushaltskundenbefragung 1991. Frankfurt.

[3] W. Ebel: Stromverbrauch in Haushalt. Institut Wohnen und Umwelt, Darmstadt 1989.

[4] W. Ebel, W. Eicke-Hennig, W. Feist, H.-M. Groscurth: Der zukünftige Heizwärinebedarf der Haushalte. Institut Wohnen und Umwelt, Darmstadt 1996.

[5] W. Ebel: Die Entwicklung des zukünftigen Strombedarfs der privaten Haushalte in der Bundesrepublik Deutschland. Institut Wohnen und Umwelt, Darmstadt 1987.

[6] W. Feist: Stromsparpotentiale bei den privaten Haushalten in der Bundesrepublik Deutschland. in: D. Sievert (Hrsg.): Zukünftiger Strombedarf Bedeutung von Einsparmöglichkeiten. TÜV Rheinland, Köln 1987.

[7] W. Feist: Passivhäuser in Mitteleuropa. Dissertation, Kassel 1993.

[8] W. Feist, Grundlagen der Gestaltung von Passivhäusem. Verlag Das Beispiel, 1996.

[9] W. Feist: Passivhaus Darmstadt Kranichstein. Bundesbaublatt, Februar 1992 (l).

[10] J. Nørgård: Husholdninger og Energi. Polyteknisk Forlag, Lyngby 1979.

[11] J. Nørgård: Improved efficiency in domestic energy use. in: Energy Policy, März 1979.

[12] J. Nørgård et al.: Development of Energy Efficient Appliances. Physics Lab. III, Technical University of Denmark, Lyngby, Denmark 1985.

[13] W. Feist: Lüftung im Passivhaus. Passivhaus-Bericht Nr. 8, Darmstadt 1994

[14] W. Feist (Hrsg): Bauvorbereitendes Forschungsprojekt Passive Häuser. Passivhaus-Ber.Nr. 2, 1992.

[15] P. Kühnle: Untersuchung der Lüftungsanlage mit Wärmerückgewinnung und Erdreichwärmetauscher im Passivhaus Darmstadt-Kranichstein. Diplomarbeit.

EFFICIENT USE OF ENERGY AT UNIVERSITIES

J. ROTURIER
C.E.N.B.G.
Université Bordeaux 1
BP 120-Le Haut Vigneau
F-33175 Gradignan Cedex, France

P. FAUCHER
Service Formation Continue,
Université Bordeaux 1
73 Avenue Roul
F-33405 Talence Cedex, France

Abstract

An important segment of the tertiary sector, namely the universities and research laboratories, usually remains outside the control of those decision-makers in charge of implementing Energy Efficiency and/or Environment Protection Programmes. In this sub-sector, the energy consumption, for example, is far from being optimised. This is either due to a lack in basic knowledge in energy efficiency or energy conservation, surprisingly often ignored by scientists or science managers, or more frequently results from a conflict of priorities in terms of decisions to be taken urgently. However, in several countries world-wide, thanks to individual initiatives or mandatory policies, Energy Efficiency and/or Environment Protection Programmes have been successfully implemented. Some of them were presented in the "ENERGY-UNIVERSITY-ENVIRONMENT" Workshop (Bordeaux - France, 1995). In the first part of this paper, a very preliminary analysis of significant case-studies for these voluntary policies and several examples from other universities are given. It is clearly demonstrated that these actions lead to significant money savings, induce new research and, more importantly, should greatly influence the training of our students as future managers concerned by these issues. In the second part, the ECOCAMPUS project, recently funded by the European Commission, is briefly presented and its main goals described. Finally, a few suggestions are given in order to analyse the advantages and/or obstacles of Energy Efficiency Programmes when implemented in the place of the standard «Laissez-Faire» policy.

S. Nørgård et al. (eds.), Development with Sustainable Use of Electricity, 205-217.

1. Introduction

About 25 years ago, a very important energy crisis resulted from the first oil shock. During more than a decade, the cost of all sources of energy was so high that managers in all areas of economic activity had to implement measures to ensure an affordable increase in the energy bill. For the same reason many new R-D programmes, aiming to identify energy-efficient technologies, were funded by governments during that period. An obvious common statement, is that such a constraint is out-of-date and that the energy issues are far from being a priority in the management of universities and research laboratories (+). Such an attitude is unaffordable in a mid-or long-range period for at least three reasons :

- the cost of oil reached increasingly higher levels up to mid-80s which was suddenly followed by a spectacular decrease. From about US$ 40 per barrel of oil, the price fell to half that by mid 96. Even taking into account the usual corrections (real versus constant money) and the significant depreciation of the US dollar compared to other currencies, it is important to remember that oil is as cheap now as in the pre-1973 period such situation being not expected to last in amid or long range period;
- in the 1970s, the biggest source of energy expenditure in most European universities was the heating sector. Moreover, since then, the electricity bill was increased significantly and has become a key concern for those in charge of university management ;
- a new source of uncertainty has progressively appeared and is now replacing the energy bill concern with the planetary aspects of an uncontrolled energy consumption with the Greenhouse effect as one of the consequences;

Despite the well-known and obvious nature of the above statement, there is today no comparable systematic support for energy-efficient R-D programmes, as there was in the mid-70s. There are many reasons to explain the present situation, the main one being the low cost of one barrel. None of these seem satisfactory. Ideally, universities and research laboratories should be perfect places for all those who work there (scientists, academic staff, engineers,...) and are actually concerned by these energy and/or environmental issues. In such places, there is no doubt that when regarding the energy/environment challenges, overall knowledge is available, i.e. the decrease in oil/natural gas resources, the nuclear waste/CO_2 debate, the growth in the demand of water or the ozone depletion. Also, when analysing the affordable development challenge, anyone can obtain the most

---

+ in the following for simplicity, instead of «universities and research laboratories» we use the single word «campus», which is given here the meaning of higher education organisations and/or private public research laboratories

recent results, analyse the possible consequences of technology development and finally reach a comprehensive view. It is hard to find a single campus where scientists are not concerned by one of these problems or any scientific organisation not supporting the need of increased R-D efforts on future energy sources such as fusion or renewable energy. An alternative approach, known as the Berkeley paradigm, has been suggested on several occasions by Pr. A.H. ROSENFELD, the *Energy Efficiency Programmes being introduced as an authentic, perfect and cheap resource* (ROS 96-a) Unfortunately, however, this is seldom regarded as one of the best ways to overcome present and future contradictions in energy policies.

University managers who pay attention to energy-efficient policies are not the majority on European campuses [1]. However, while preparing the present paper, we have been able to identify such campus sites where, as a consequence of national or local policies, managers who are seriously concerned by these issues do exist. Not only have they decided to implement an electricity tariff policy, a very popular solution, but they give top priority to energy efficiency or environment protection criteria in <u>every investment decision</u>. Among other benefits, such programmes result in a significant decrease in the energy (or water demand [2]) leading to a non-negligible reduction in related bills. However, up to now, in spite of these advantages, a limited number of campuses only are involved. That seems paradoxical but may be understood, as underlined by R.H. SOCOLOW (SOC 94), ... *"Energy efficiency which today is so often rejected as uninteresting and confining, <u>will become</u> embedded in human culture"*. This has not yet happened and, on a typical campus, from the energy efficiency and/or environment protection point of view, most users or managers do not behave any differently to typical end-users in other parts of the economy.

The present study was started in 1992 and we have met since then many colleagues across Europe and have established contacts world-wide. From that first and limited experience, we are convinced that, concerning energy efficiency and/or environment protection issues, the typical scientist's workplace is simply a "TERRA INCOGNITA". However, we have been able to note some moves to a more conscious attitude, more and more often based on individual initiative or regional/national programmes. Mainly based on such success stories, our main objective in the present work is a contribution to let the scientific community know the numerous benefits of a synergy effect when Clean Production goals are associated to each step of the «Scientific Process». In a preliminary survey of the presently available information, we firstly summarise the main results presented to the "ENERGY-UNIVERSITY-ENVIRONMENT" [3] Workshop : the most important benefits and results from a few case-studies are given, a few obstacles being

---

[1] the same being possibly more or less true for more environmentally sound policies such as in water use or waste management programmes

[2] Our work includes the environment protection issues (i.e. water demand, waste production) as a key issue. However, a detailed presentation of such programmes is beyond the scope of the present paper

[3] cited in the following as EUE-95

underlined too. An example of initiatives resulting from EUE-95, the ECOCAMPUS action, a European network of campuses concerned by Energy-Efficient and Environment-Safe Management, presently being operated (ECO 96) is briefly introduced in a second part. Finally, a few preliminary results obtained over recent years in the University of Bordeaux 1 at Talence are shown as a case-study.

2. Energy-University-Environment Declaration Statement (EUE-95)

The "EUE-95" Workshop, hosted by the University of Bordeaux (France) in March 1995, was mainly supported by the European Commission Directorate General for Environment. About 50 participants from 12 countries recognised that in most of their workplaces, the management of natural resources, including energy and/or water use, is usually based on a "Laissez-Faire" policy. That appeared to be an unwise, unfair and unaffordable situation and, as a first action, all participants then decided to vote a Declaration Statement [4] of which extracts are given in the present paper (printed in italics).

First of all, the main three reasons to discontinue the present standard "Laissez-faire" Policy and a few success stories are listed as below:

> *a)- The financial cost associated with the daily operation of equipment to support our teaching or research activity continues to grow as more electronics are added to our laboratories and as more computers are added to our offices. The environmental degradation associated with the power production for the operation of heating, cooling (including cooling water for science equipment such as lasers, etc.) and lighting to support teaching and research activities has also grown as energy use has grown.*
>
> *b)- The needs for the development of new research and development programs are not being addressed in all countries around the world. This is due in part to artificially low energy prices and a lack of public awareness about the uncertainty of future energy supplies. Compounding this problem is the fact that research results are published in a multitude of obscure journals, often in a language foreign to the researcher. Hence, results from existing programs are not quickly distributed to others around the world. Furthermore, specific details about experimental set-ups and assumptions are often buried in unpublished reports and theses that can take years to discover.*
>
> *c)- Our students are not receiving the best training because of limited, out-of-date texts, bad examples in campus buildings and isolation from others around the world who are interested in efficient building design and protection of the environment.*

---

[4] a complete version, in English or in French , is available. (Also translated into German and Polish)

As previously indicated, the management of university campus and research laboratories frequently ignore the Energy Efficiency and/or Environment Protection Programmes. For scientists or science managers there is no priority nor, even, a standard requirement in these issues. In such conditions, the typical energy management programme is implemented on a "Laissez-faire" basis, most of decision-makers within the scientific community having no idea of the benefits resulting from the simple Demand-Side Management (DSM) or the more complete Integrated Resource Programmes (IRP) (ALM 93). However, their benefits having been clearly demonstrated in many case-studies from industry or the commercial sector, based on individual initiatives or mandatory policies, DSM/IRP programmes are successfully operated in universities and research laboratories world-wide. Some of these were presented in the EUE-95 Workshop and associated here to several examples from other universities identified since then, a very preliminary analysis of such significant case-studies [5] being given. These actions lead to significant money savings, stimulate new research areas and, probably more importantly, should greatly influence the training of students considered as future managers inevitably concerned by these issues:

a)-The $98.6 million Texas LoanSTAR program, conducted by Texas A&M University in College Station, Texas, through funding supplied by the Texas State Energy Conservation Office, has achieved 120-150 % of audit estimated savings by carefully measuring before/after whole-building energy use and using the measured data to find and fix operational and maintenance problems.

In the last 4 years investments of $53 million have yielded measured retrofit savings of $16 million per year. Actually, achieving such a success story requires (CLA 96) :

* a very careful measuring campaign before/after whole-building energy use
* the use of the measured data to find and fix operational and maintenance problems
* a strong involvement of several researchers

b)- The Asian Institute of Technology in Bangkok, Thailand, where conservation policy allows students to share in the cost savings through the purchase of new books for the library.

c)- ETH, the Swiss Federal Institute of Technology, has recently launched on its Zurich campus an energy conservation program in which co-operation between users and building-maintenance staff is systematically improved. Professors from several departments (architecture, civil engineering and electrical engineering) are engaged in the retrofit of campus buildings. The entire staff is invited to

[5] similar policies are now being implemented for Water Management such as on the INSA Campus in Rennes, France, where a closed-loop water cooling circuit is in operation decreasing by at least 40 % the water demand.

participate in a group-oriented, self-learning program that aims at activating everybody to take energy conservation measures in the labs and offices, as well as in the orientation of teaching and research.
d)- In Manchester University in the United Kingdom, energy consumption has been cut by 36 % over the last 6 years, thereby achieving an annual savings equivalent to the pay of over 40 teaching staff.
e)- The University of Coimbra in Portugal has carried out a comprehensive audit to identify the opportunities for electricity conservation, retrofit measures being expected in the near future.

Other examples were recently pointed out. In the Technical University of GDANSK (Poland), a study of the energy conservation programme was started in 1994 (KAM 95). In the University of AUCKLAND (New Zealand), the managers were faced with the continuous growth in electricity consumption which seems to be unavoidable to most of us too! Water heating, space heating and cooling, and lighting systems, have been finely tuned over the last 15 years. The most significant result is shown through the electricity use ratio calculated have dropped from 1700 kWh per student in 1988 to 900 kWh per student in 1995, a decrease by a factor of two !

In two EU countries, these programmes should hopefully become a standard as a result of two recent nation-wide initiatives, in the UK by the Council of Vice Chancellors and Principals (CVCP) and in Austria by the Austrian Agency for Energy Management (EVA). The CVCP has recently published (STR 96) a short note underlining the vital role of the educational sector as a major requirement «... in encouraging better environmental management (i) through example (ii) through the provision of the knowledge and expertise... ». British Universities are therefore recognised as having «... the potential to make an enormous contribution to this Process but, as yet, few have successfully incorporated environmental concerns into their own institutional practices ... ». In order to disseminate as soon as possible, a list of «Good practice», CVCP intends to publish the report «Environmental Management in Universities» intending to give an appropriate reply to important questions i.e. : How to implement Environmental Management ? How to develop «Best Practice» (energy end-use, waste, building management). Such an information package should focus, in particular, on (i) an overview (international, national), (ii) the most significant data outlining both the scale of the problem and the potential contributions of campus, (iii) case-studies and sources of information. In Austria, the Programme :«Energieeffiziente Universitaeten», now understudy should be mandatory on a nation-wide basis (BEN 96). One of the main characteristics of that programme is the (expected) strong participation of students as in the Technical University of Vienna, where a preliminary work is being performed with a group of 12 students.

Let us conclude this first part with a new, exciting, idea suggesting a possible improvement of the energy and water management in CERN, the European Particle

Physics Laboratory in Geneva (Switzerland). Annually CERN consumes 1,000 GWh of electricity, about 2 % of the Swiss total electricity consumption, and 25 million m^3 of water, mainly to cool the equipment that induces about 800 GWh of wasted thermal energy, a large part of which may be, at least from a technical point of view, reused. As part of his Master thesis in Energy submitted in 1995 to the Swiss Federal Institute of Lausanne (EPFL), W. Van SPROLANT (SPR 94), has discussed the feasibility of an entirely new project aiming to recover, in part or in full, the very large amount of calories embedded in the water used to cool the equipment. Two different options are investigated, all consequences being deeply analysed, in particular the technology and economics issues, not ignoring related problems such as the compliance of the reused water with the Swiss standards. In a first, and probably unrealistic, option, it is assumed that CERN should shift its normal activity from summer to winter i.e. scheduling the use of the accelerators and associated experiments. In such situation, about 800 GWh of the thermal, presently wasted energy, should be recovered. The calories are extracted through heat exchangers, the temperature of the water to be reused is about 30 °C. The Van SROLANT's CORSAIRE project [6] is nothing but an opportunity to make possible a wintertime correction of the public drinking water network. Presently delivered by the water utility at a low 5-6 °C temperature and circulating in pipes along the walls, it results in a freezing of the buildings. Would the CORSAIRE project be achieved, the water temperature is about 15-20 °C and such freezing obviously non-existing. Unsurprisingly, the total primary energy savings and avoided power associated to that opportunity are far from being negligible. In particular, the peak or semi-peak demand should be reduced not ignoring that the energy savings are equivalent to a decrease of 190 kT of CO_2 each year. Annual extra benefits are demonstrated for the CERN too : (i) a 40 GWh decrease of its own electricity consumption (ii) a shutdown of the present cooling circuit and towers which consume 20 millions of m^3 of water (30 % of the consumption of the City of Geneva). However, presently, such Summer to Winter shift seems to be unlikely, a second option is proposed. The calories generated in Summer are stored in a very large insulated and non-rigid tank open on its lower part and built in the bottom of Lake of Geneva. In wintertime, the water should be pumped and used as previously shown in the CORSAIRE project. Actually, in spite of its large conservation potential, such project is to remain «on-the-shelf», at least in CERN in a mid-range perspective. However, such data were successfully extrapolated by Van SPROLANT's to industrial plants, i.e. power plants generating a large amount of wasted warm water (Table 2) [7]. As confirmed by A.H. ROSENFELD (ROS 96-b), that seems to have become a new option available to policy-makers..

---

[6] CORSAIRE : CORrection SAIsonnière de la température du Réseau d'Eau potable

[7] Recently, feasibility studies part of the CORSAIRE Proposal, have been regarded by officials in the Swiss Neuchatel and Geneva Cantons

Table 1. Van SPROLANT's extrapolated data to Switzerland and the whole EU

	Switzerland	EU
Primary energy conservation potential	**0.7 MTEP/yr**	**36 MTEP/yr (*)**
Avoided Peak Power	**340 MW**	**18 GW**
Money Savings	**$ 800 M/yr**	
Avoided CO^2 emissions	**3 MT/yr**	**120 MT/yr (**)**

(*) equivalent to the electricity generated in one year in France
(**) as calculated by the authors of the present paper

3. The ECOCAMPUS Programme

Several participants to the EUE Seminar have decided to implement a follow-up by creating a Work Group targeting two main goals: (i) data collection and analysis (ii) identification of the obstacles. They were also expecting to perform preliminary studies in their own organisation in order to analyse the possibility of disseminating the results to the whole scientific community. The EU-DG XVII JOULE / THERMIE B Programme appearing to be a perfect opportunity for that purpose, a proposal was submitted by 9 of them, scientists or consultants, working in institutes or companies from 5 EU and 2 Eastern European countries:

* Technical University of Denmark
* Olof Granlund Oy (Finland)
* University of Bordeaux (France)
* Technical University of KEVALA (Greece)
* Univ. of Mining and Metallurgy of KRAKOW (Poland)
* University of COIMBRA (Portugal)
* Institute of Building Engineering of BUCHAREST (Rumania)

The work programme, accepted in late 1996 and scheduled to last 13 months, aims to:

- create a European network of University campuses and Research laboratories where managers and/or Academic Staff or scientists and/or engineers are interested in establishing Clean Production as a key priority. Regarding the energy issues, in particular, any investment decision such as retrofit of ancient or construction of new buildings, construction of new research facilities, purchase of equipment would identify the efficient-use of energy as one of the important criteria
- collect and analyse all data linked to energy use and/or environment protection in investment and operation of all campus facilities

- define a specific methodology to manage such very special sites as University campuses and Research laboratories and perform a certain number of feasibility studies
- demonstrate the benefits of a synergy effect when the Clean Production issues, that include Energy Efficiency Programmes, are linked to each step of the «Scientific Process», not ignoring, however, an analysis of the obstacles and constraints
- organise the different steps in such conditions that the students can actually participate as part of their own courses or programmes.

One of the first steps, initiated when preparing the EUE Conference, is to establish a database including the main specifications in the energy sector [8], for each Institute or University campus regarded as an energy/water end-user and waste producer. To our knowledge, such data banks having not yet been created, in the present state-of-the-art, it is impossible to accurately compare real data between campuses with similar number of students and research activities. Such figures could be useful in studies, such as those performed in the framework of the IMHE/OECD (OEC 96). In that work, for example, the budget of several universities world-wide is investigated and the available resources quite accurately given. However, the operating costs are not systematically shown, nor are the sums represented by these costs in the large bills corresponding to energy and water. Were such data to become available, there is no doubt that it would very easily demonstrate that energy efficiency programmes resulting in large cost savings, have to be developed. Translated into a significant reduction in the cost of providing a quality education, policy-makers would certainly be interested in such important when considering costs in the 90's.

4. The University of Bordeaux 1 Case Study

To-day, nobody can deny the huge increase in electricity consumption that results from the growing use of Information and Communication Technologies (I&CTs). More and more Computer Engineering Departments or Computer Centres have to face significant increases in the resulting electricity bills. This issue has been analysed by the OT3E Work Group funded by the EU DG XVII and chaired by University of Bordeaux [9]. Some interesting data summarises the effect of the growth of I&CTs across Europe:

- their total present power load is about 15-20 GW;
- the annual growth of the electricity demand resulting from new equipment is equivalent to 2 large 1,000 MW power plants

---

[8] Other area associated to the «Clean Production» issues such as water demand and waste production are part of the study

[9] see in the present proceedings, the paper by J.ROTURIER and A. ANGLADE «Energy-Efficient Information and Communication Technologies» ... »

- techniques to limit electricity consumption have been built in almost all new Office Equipment (PCs, monitors, ..) supplied to European customers for over 2 years. However, only 10-20 % of users are commonly using the «Energy Star» energy-efficient device in their own central units or monitors
- a better user-friendly approach has not yet been achieved in particular for PCs and Monitors

As long as the energy conservation potential of the crucial I&CTs sector is being assessed, the US-EPA [10] Energy Star Programme must be taken into consideration also demonstrated as a perfect case-study to illustrate the constraints the users are facing with when intending to establish the related resource in Rosenfeld's definition (see Introduction). Actually this energy efficiency programme, possibly allowing important benefits typically 50 % of the total Office Equipment electricity consumption, is implemented by a very limited number of users, scientists and academic staff being not more concerned than other people.

In many other electricity end-use sectors (lighting, air-conditioning, ventilation, heating and cooling), a conservation potential may be calculated using the standard procedures alike in any other office buildings. From our present (and limited) data using common parameters, we easily demonstrate the cost-effectiveness of using energy efficient techniques. In order to provide data to support a possible feasibility study, a preliminary work has been now underway, in the University of Bordeaux 1, for several years. Interesting data regarding the electricity demand and water consumption in the 1986-96 decade are shown in Table 2.

Table 2. Evolution of the electricity and water consumption in 10 years (University of Bordeaux 1 : **1986-** 1996)

		19 86			1996	
Electricity consumption	**GWh**	**kWh/m²**	**kWh/stud**	**GWh**	**kWh/m²**	**kWh/stu d**
	6.87	**62**	**686**	**10.06**	**70**	**940**
Water consumption	**thous.m^3**	**m^3/m²**	**m^3/stud.**	**thous.m^3**	**m^3/m²**	**m^3/stud.**
	640	**5.8**	**64**	**526**	**3.66**	**49**

Table 2 suggest two important remarks. When the electricity consumption is compared to similar data from the University of Auckland (see p. 5), a significant difference is demonstrated with a decrease by a factor of 2 from 1988 to 1995 in the latter University and a growth of the kWh/student ratio by about 30 % in the Bordeaux case-study. Regarding the water consumption in Bordeaux, both m^3/m² and m^3/student indicators show another important decrease. That can be probably be explained by the nature of the

[10] Environment Protection Agency

water resource of which a large part is found locally, the complement, about 10 %, being provided by the Bordeaux water utility. The local resource, in particular due to its much lower cost, is so important that it is carefully managed and regarded in priority in any new investment decision by the technical staff. Such comments, clearly demonstratedthe advantages of a voluntary policy when compared to a «Laissez-Faire» Policy.

5. Conclusions

As previously underlined in the EUE-Declaration Statement, we suggest keeping in mind certain issues when intending to implement, as our US colleagues say, a "Greening of the Campus" Programme. First of all significant barriers must be identified and removed, before such programmes can be adopted, the most important of them being shown below :

1. *The lack of social concern for reducing energy use and the associated environmental pollutants*
2. *Low priority for energy and environmental matters due to the relatively moderate cost of the heating/cooling/lighting of buildings*
3. *Insufficient or inadequate capital funds for making conservation investments*
4. *Inappropriate financial mechanisms for capturing energy savings in accounts that can be used to finance additional conservation investments*
5. *In countries in economic transition, other difficulties also appear including :*
    * *unclear relations between sustainable development principles, economic constraints and energy-efficient or environment-safe policies,*
    * *divergent economic policy goals such as rapid growth and dwindling resources*
    * *inadequate and ineffective privatisation of state property.*

Moreover, participants such as scientists and teachers actively involved in energy efficiency, protection of the environment and economic development, should not be satisfied with simply publishing their EUE-95 Declaration of Statement. Each of them also feel the need to actively demonstrate on one's own campus or in one's own laboratory and commit oneself to launching initiatives such as to:

- summarise and communicate the results of the 1995 EUE seminar to the administrative boards of our institutions and to our fellow colleagues particularly those who have responsibility for the management of the campus and facilities ;
- identify, within its own institutions, such sites where energy and/or conservation opportunities are feasible ;
- provide a positive influence on the campus purchases of energy consuming equipment by helping decision makers purchase efficient equipment ;

- promote the fundamentals of efficient use of energy and water in all courses that are taught, together with the protection of the environmental and sustained economic prosperity ;
- post the results of these efforts in a public domain access site (i.e. an Internet bulletin board) where others can learn quickly from their experiences, and participate with experiences of their own that will serve as an efficient, informative model for all people to learn from.

More generally, an important priority is to convince all decision-makers concerned with energy efficiency and/or environment protection policies to *pay a very close attention to the education of future decision makers*. Such a policy when initiated and sustained by an adequately supported Programme, such as ECOCAMPUS, will reward policy makers by reducing education costs through reduced campus costs (heating, cooling, lighting, ...). Actually, although the present «Laissez-Faire» Policies become clearly unaffordable, the scientific community remains poorly concerned. So, finally, we feel useful to possibly try to understand, explain and override that situation. However, our own experience in that area is limited to the France case-study and comments provided by EUE-95 Seminar participants. We suggest that to analyse, on a much broader sample, all those obstacles and/or constraints which prevent people from moving from the standard "supply-oriented" energy training to a more appropriate "demand-side oriented" approach.

6. Refrences

ALM 93 A. de ALMEIDA «An Introduction to Integrated Resource Planning» in INTEGRATED ELECTRICITY RESOURCE PLANNING - Proc. of the 1993 NATO - ARW Espinho (Portugal) edit. by A. de ALMEIDA, A.H. ROSENFELD, J. ROTURIER, J. NORGARD - Kluwer Academic Press 1994.

CLA 96 D.E. CLARIDGE «Processes Used in Performing Successful Energy Audits» IMPROVING ENERGY EFFICIENCY OF COMMERCIAL BUILDINGS WORKSHOP - United Nations Bangkok Thailand (Nov. 1996).

ECO 95 Contract CEC-DG XI # B7 - 81110/94/282/AO/B4.

ECO 96 ECOCAMPUS EC-DG XVII THERMIE Programme Contract # STR 1006 96 FR.

EUE-95 Energy-University-Environment Seminar (University of Bordeaux, 1995) Declaration Statement (Contact J. ROTURIER).

KAM 95 W. KAMRAT, A. AUGUSIAK «Selected Problems of Energy Savings Evaluation in University Buildings» Private Communication.

ROS 96a A.H. ROSENFELD «Climate Changes : Status and Risks» and «Implementing Efficiency in USA» in the present NATO-ARW Proceedings.

ROS 96b A.H. ROSENFELD Private Communication.

ROT 95 J. ROTURIER A.H. ROSENFELD "Energy Conservation Starts in the Lab" - Physics World Dec. 1995.

SOC 93 R.H. SOCOLOW in "USING ENERGY IN AN INTELLIGENT WAY" - Proc. 11th WE-HERAEUS SEMINAR - ed. by E.W.LINGEMAN - European Physical Society Geneva 1993.

OEC 96 OECD-IMHE Programme (see for example «Gestion de l'Enseignement Supérieur Vol. 8, # 1, 1996»

SPR 94 W. Van SPROLANT «Correction saisonnnière de la température du réseau d'eau potable du canton de Genève par la valorisation des rejets thermiques du Laboratoire Européen pour la Physique des Particules (CERN)» - M.Sc. Thesis Ecole Polytechnique Fédérale de LAUSANNE (Juin 1995).

STR 96 V. STRANG Private Communication.

LEAST-COST PLANNING ON ELECTRICITY

STEFAN THOMAS
Wuppertal Institute for Climate Environment Energy
Döppersberg 19
D-42103 Wuppertal, Germany

1. Introduction

During the last years, an increasing number of utilities in several European countries, e.g., Denmark, France, Germany, Ireland, The Netherlands, Portugal, Spain, and Sweden, have discovered that their business can go beyond selling kilowatt-hours of electricity or gas. In a recent survey by VDEW, the association of German electric utilities, around 80 utilities from all levels (there are nine connected grid companies generating more than 80 % of all power, around 50 regional and about 800 municipal electricity distribution utilities in Germany; most of the municipal utilities offer also natural gas and sometimes district heat from cogeneration plants) answered that they offered over 200 demand-side management (DSM) programmes to their customers (VDEW 1994).

Why do utilities engage on the demand side of the market to reduce electricity consumption and peak demand? There are two main reasons why utilities should do so:
First, it is cheaper to conserve energy than to produce it. The reason for this is that "markets" fail to achieve the optimal mix of efficient end-use technology and energy supply.

And second, the threat of catastrophic climate change imposes on industrialised countries the need to reduce their CO_2 emissions by 80 to 90 % until the middle of the next century. In the short and medium term, saving fossil fuels by efficient end-use of energy can make the biggest contribution to this emissions reduction.
Therefore, the most effective way to reduce environmental problems is also beneficial to the economy.

Since 1980, new rational planning techniques for electric and gas utilities have been developed, and have increasingly been used in North America and Europe which aim to include the rational use of energy into the utilities' resource portfolio and strategic planning. These techniques are widely known as Integrated Resource Planning (IRP) or

S. Nørgård et al. (eds.), Development with Sustainable Use of Electricity, 219-236.

Least-Cost Planning (LCP), because they seek to integrate the least-cost options from both the supply side and the demand side into a utility's business plan[1].

In the following, I will discuss the principle of Integrated resource planning, its use in strategic utility planning, and its economic implications for the utility and its customers. These issues will be clarified with the results of one of the most comprehensive IRP processes completed so far in Europe, the LCP case study which was prepared with co-funding from the EU's SAVE programme for Stadtwerke Hannover, a German municipal utility serving about 300.000 customers (Öko-Institute / Wuppertal Institute 1995).

2. The IRP principle

Integrated resource planning can be used by electric as well as gas utilities, but for reasons of simplicity I concentrate on the electricity sector. The main differences between IRP and traditional utility planning are:

1. *Energy services* (e.g., properly lit and heated rooms, "hot pizza and cold beer"), not energy (i.e., kWh) have to be provided with as little damage to health and environment and as cheap as possible. Energy is no longer the end product, but only an intermediate product. This may include new, dispersed generation from small-scale cogeneration and renewables, but the most important new idea within the concept of energy services is that
2. Utilities engage to *reduce energy consumption* whenever this is cheaper than providing additional energy ("demand-side management", "the business behind the meter"). Ideally, the utility integrates suppliers of energy-efficient end-use technology and an appropriate form of energy supply into an efficient energy service package offered to the customer.
3. The *level playing field* between demand-side and supply-side resources means that both options should be analysed using the same economic methodology and parameters, especially regarding pay-back times and interest rates.

---

1 Thus, the term "planning" does not imply that state authorities are to impose plans onto utilities. Instead, IRP is understood as an extension of traditional utility business planning to integrate new types of resources on the demand side ("the business behind the meter") and from new, especially dispersed sources of generation (e.g., small-scale CHP and new renewables). The state, however, has to provide the framework which makes it legally and economically viable for utilities to include all types of resources into their business plan which have environmental and net economic benefits to society.

The first step in identifying the demand-side energy efficiency potential is to examine in detail the actual end-use consumption as well as the technical possibilities and extra costs for reducing this consumption through more efficient end-use technology. In the Hanover example, this was done for 10 electric end-uses each in the residential, commercial/public, and industrial sectors; both end-use energy efficiency and fuel substitution technologies were examined. Fuel efficiency options for 5 end-uses were also analysed. The results are expressed as savings potential (in GWh/year) relative to a "business-as-usual" technology, and as the cost of conserved energy (in Pf/kWh). The different options can be ordered by increasing cost of conserved energy to give a "supply-curve of demand-side energy efficiency", as shown in Fig. 1 for the Hanover example.

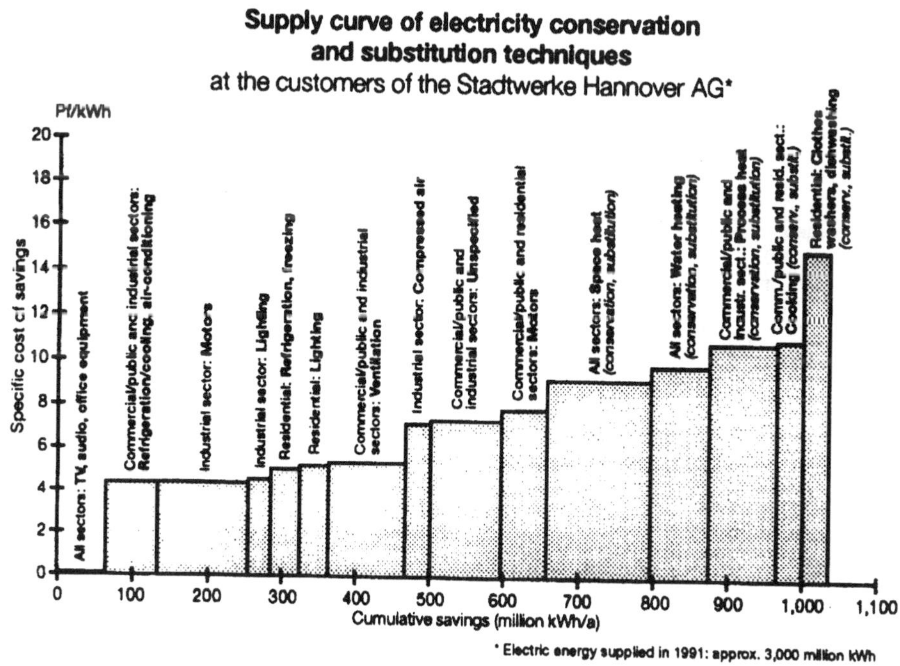

Figure 1.
Supply curve of conservable or substitutable electricity for Stadtwerke Hannover (consumption in 1991: 3.027 GWh).

Fig. 1 shows the accumulated savings to the right and the costs of conserved energy to the top. Thus it is possible to detect the biggest and cheapest savings potentials which are most promising to address with DSM programmes.

A similar supply curve was also developed for savings in non-electricity fuels. The following table summarises the results.

TABLE 1. Economic potential for energy savings in Stadtwerke Hannover's electricity service area.

Energy	Electricity		Gas, Oil, District Heat, Coal	
Sector	Consumption in 1991 (GWh)	Reduction potential	Consumption in 1991 (GWh)	Reduction potential
Industry	1.118	30 %	2.931	19 %
Commerce/Public	1.134	35 %	3.312	28 %
Residential	775	37 %	3.931	34 %
Total	**3.027**	**34 %**	**10.174**	**28 %**

About one third of all end-use energy consumed in Hanover could be conserved (cf. Table 1 and Fig. 1) at lower costs than the long-run marginal costs of electricity supply (including avoided fixed and variable costs of generation, avoided losses, and avoidable costs of electricity transport, distribution, and sale), which are between 0,07 and 0,086 ECU/kWh (13 and 16 Pf/kWh) without externalities for the Hanover case, respectively at costs lower than the marginal costs of non-electricity fuel supply which are between 0,005 and 0,022 ECU/kWh (1 to 4 Pf/kWh).

Why do customers not realise these enormous cost-effective savings potentials on their own? Because there are a large number of *market barriers* and failures on the demand side. Among the most important of these are:

1. Lack of knowledge: energy users, especially small- and medium-size companies and private households, normally are not professionals in the efficient use of energy, whereas the supply of energy has been a professional business for 100 years. To those industrial companies, municipalities, and institutions which have installed energy managers this should no longer pose a problem.
2. Suppliers of end-use technologies often have economic disincentives to offer the most energy-efficient options to their customers (e.g., when engineers do not get paid for extra planning time which would be needed for designing more efficient lighting or ventilation facilities).

3. Quite often, the user of a building or plant is not the owner. In these cases, e.g. for rented office space, the owner seeks to minimise investment in energy-consuming equipment, since he will be paid the energy costs from the user. Thus, the user has an incentive to reduce energy costs but usually no influence on the investment. It is not possible for the user to choose an equipment which has higher investment costs but lower energy consumption.
4. Public entities often have tight budgets and face problems to invest in energy efficiency. Furthermore, the rules of public budgeting can make it difficult for them to finance energy efficiency investments from savings in running costs.
5. The pay-back gap: Most industrial energy consumers apply the same investment criteria for energy efficiency investments as for investments in their core business, e.g. for new production lines. Thus they require pay-back times of three years or less. While this may be a prudent behaviour from the perspective of an individual company, it leads to systematic over-investment in energy supply. Utilities generally apply pay-back periods of 15 to 20 years to their investments.

These market barriers and imperfections are the reason why utilities should engage to realise the cost-effective energy saving potentials. Especially the pay-back gap makes it necessary to create a level playing field between supply side and demand side through IRP.

An example can make these issues clearer. It is taken from one of six pilot studies of industrial, commercial, and public customers conducted within the Hanover LCP case study. In an office building of an insurance company, we found that renewal of the lighting system could save 30 % of the electricity that had been used so far for lighting. The old lighting system had to be renewed anyway; but half of the savings from the new system could only be achieved by using electronic high frequency ballasts instead of conventional low-loss ballasts.

The following Table 2 shows how costs of conserved energy are calculated for this example. The additional investment for high-frequency ballasts was 35.000 DM out of a total of 81.600 DM. Only this extra investment may be taken into account for calculating the costs of the electricity conserved through high-frequency vs. conventional ballasts. The level playing field is created by calculating the levelised life cycle costs with a societal real discount rate of 4 % (7 % nominal minus 3 % inflation rate) and for the useful life of 15 years of the installation, yielding a levelisation factor of 0,09. The costs of conserved energy are then calculated by dividing the levelised life cycle costs through the electricity conserved.

From the society's perspective, this investment is clearly cost-effective with costs of conserved energy of 8 Pf/kWh, as compared to avoided electricity supply costs of 13 or 16 Pf/kWh (s. above).

TABLE 2. The pay-back gap: Costs of conserved energy from society's and user's perspective

	society's perspective	user's perspective
total investment	81.600 DM	81.600 DM
extra investment for energy efficiency	35.000 DM	35.000 DM
useful life	15 y	15 y
required pay-back time	15 y	3 y
required discount rate	4 %/y	6 %/y
levelisation factor	0,09	0,37
levelised life cycle costs	3.150 DM/y	13.090 DM/y
electricity conserved	39.700 kWh/y	39.700 kWh/y
costs of conserved energy	**7,9 Pf/kWh**	**33,0 Pf/kWh**

Furthermore, the right column in Table 2 gives an example for the pay-back gap. Consider a company requiring three years pay-back for the investment. This has the same effect as if the company required a real discount rate of 39 %. The levelisation factor becomes 0,39, the costs of conserved energy *from the customer's perspective* are almost 35 Pf/kWh, which is above the consumer's electricity price of 20 to 25 Pf/kWh (typical for medium-sized commercial customers).

By offering a rebate of about 30 to 50 % of the extra cost, the utility could reduce the costs of conserved energy from the customer's perspective below the electricity price (i.e., the pay-back period below three years) and make the investment attractive to the customer. Or the utility might offer a third-party financing scheme, where the utility initially pays the investment to the customer and then receives monthly payments proportional to the predicted or measured savings for, e.g., five or six years, until the investment and an appropriate return have been paid by the customer.

3. IRP as an instrument of strategic utility planning

As shown in the previous section, the existence of cost-effective energy efficiency potentials is the reason and the starting point for making IRP an instrument of strategic utility planning. Whenever end-use energy efficiency is cheaper than production and supply of end-use energy, it should be made use of as a resource ("NEGAWatt"). Utilities should encourage and support consumers to use energy-efficient end-use technologies in order to achieve lower overall costs of energy services. By doing so, utilities should try to achieve quantitative targets of conserved energy and capacity, which are determined through a cost-benefit analysis during an IRP process. This means going beyond the general information and advice utilities have been giving to their customers, e.g. through customer advice centres.

There are five ways to deliver such targeted support in DSM activities and programmes targeting specific customer sectors and end-use technologies:

1. *Consulting*. By giving energy-audits at the customer's premises free of charge or at a low fee, utilities can help customers detect cost-effective energy efficiency potentials. This can also be done by consultants which are paid for by the utilities. Danish utilities have conducted or paid more than 4.000 audits in industrial and commercial companies, and public institutions during the last years. On average, about 10 % savings potential with an average pay-back of two years have been detected, of which 45 % have been implemented by customers (Elser 1994, ELSAM 1994).
2. The rate of implementation of energy efficiency measures detected during an audit may be increased if a *rebate* on the investment is offered together with the audit. This may be a fixed amount for standard devices like high-frequency ballasts. Such fixed rebates are also very common among German utilities for electric household appliances. RWE Energie, the largest German utility, paid one million rebates of 100 DM each for efficient refrigerators, freezers, dish-washers, and washing machines between 1992 and 1995. The programme saved 450 GWh of electricity; the costs to society were below the long-run marginal costs of electricity supply (Wuppertal Institute / BEM 1995, Thomas 1995). For larger investments in individual technical solutions, a rebate based on the predicted yearly savings is appropriate, e.g. the "customized rebate" programme by the largest Californian utility Pacific Gas & Electric.
3. With *direct installation* or free give-away programmes, utilities offer comparatively cheap devices like compact fluorescent lamps (CFLs) or low-flow shower heads to their customers for free. These programmes can be very effective, reaching high participation rates, and also often rather cheap. Many European utilities have offered CFLs free to their residential, and sometimes small commercial customers (Mills 1991). One example is the programme Stadtwerke Saarbrücken, a municipal utility serving 100.000 customers, implemented in 1994. More than 50 % of the residential customers came to the advice centre of the utility to get their free lamp. About 2 million kWh were saved cost-effectively to the utility and the society. This success was possible because the offer of the free lamp was combined with an advertising campaign making people familiar with CFLs. Utilities in the USA are also offering such free services to small commercial customers.

4. An even bigger financial engagement is required with *third-party financing*. On the other hand, third-party financing allows the utility to recover its expenditures directly from the individual customer via bilateral contracts, while expenditures for other types of programmes with audits and/or financial incentives have to be recovered through rates from all customers. Many European utilities are examining ways to expand their business to the demand side and introduce energy-efficient services through third-party financing, and there are now a lot of pilot schemes under way.
5. An even more innovative and market-oriented possibility to include also large customers and private energy service companies into the efforts to realise cost-effective electricity savings is *demand-side bidding*. In DSM bidding schemes, the utility publishes the need for saved capacity and electricity and asks for offers. Large industrial customers or consultants who have the technical skills to develop projects for electricity efficiency can offer these projects to the utility and ask for a certain payment per certified kW of kWh from the utility. The utility chooses from the projects up to a certain limit of capacity or a certain price cap and makes contracts with the successful bidders to provide and proof the savings. DSM bidding has become increasingly common in the USA during the last years (Goldman/Kito 1994) but has not yet been tried in Europe.

Not every activity connected with DSM programmes of these type has to be carried out by the utility itself. The utility will probably have neither the capability nor the capacity to perform a large number of energy audits on all technical details of efficient electric end-use technology. The utility, however, should provide the financial and organisational framework, thus creating an opportunity and a market for engineering consultants, technical contractors, and customers to carry out more energy efficiency measures than they would do without DSM programmes.

4. Development of nine prototype DSM programmes

For Stadtwerke Hannover, we designed nine DSM programmes targeting all customer groups and those end-uses with the biggest savings potential. Fig 2 displays the customer groups and end-uses as well as the type of programme chosen. During the LCP case study, there were also two pilot schemes: one for third-party financing with six pilot studies of industrial, commercial, and public customers already mentioned above, and one with rebates for efficient refrigerators and freezers for residential customers.

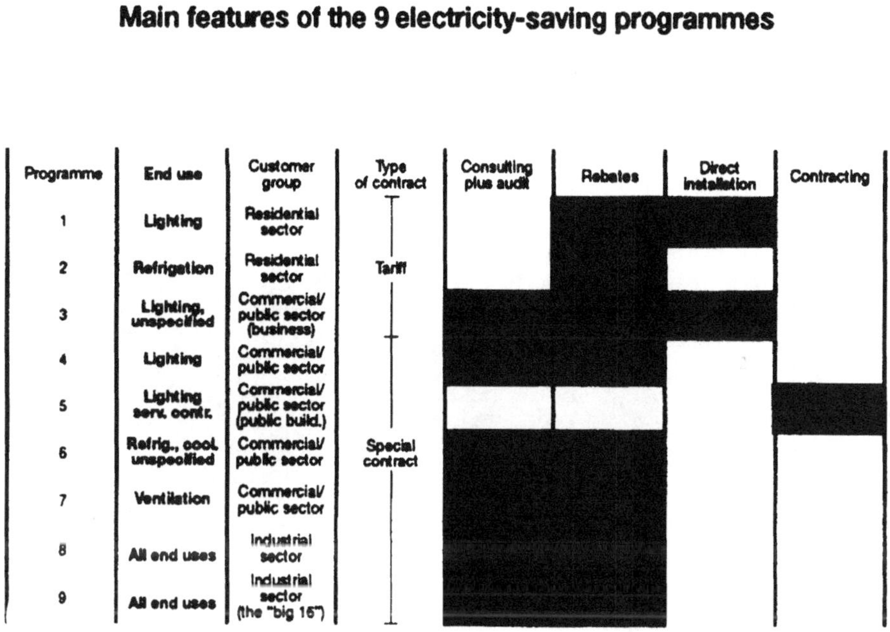

Figure 2.
Nine DSM programmes for Stadtwerke Hannover.

Based on these experiences and on a discussion with the utility about the expectable acceptance and participation by customers of these programmes, and using specific load curves of conserved electricity for each programme, we calculated the expected capacity and energy savings from these nine programmes. They add up to about 40 MW of capacity and 200 GWh of electricity after 7 years of programmes (Fig. 3). The reason why savings depicted in Fig. 3 decline after the year 2005 is that more and more of the energy-efficient appliances and installations reach the end of their useful service life. This trend could well be overcome by implementing follow-up and additional programmes after the year 2000. Thus the savings relative to the "business-as-usual" case could reach even more than the 6 or 7 % expected from the nine prototype programmes. The favourable boundary conditions mentioned in Fig. 3 will be addressed below.

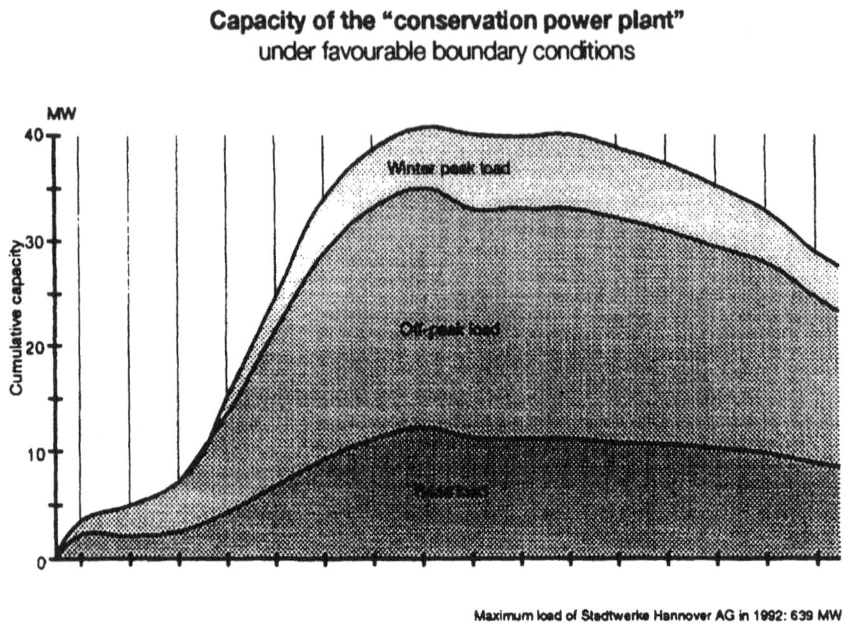

Figure 3.
Savings from nine DSM programmes for Stadtwerke Hannover between 1995 and 2010.

Almost 90 % of the savings from these nine DSM programmes come from the commercial and industrial sector, and their benefit-to-cost ratio is predicted to be rather good (Fig. 4): The benefits, i.e. the long-term avoided marginal costs of electricity supply, are forecast to be higher than the costs, i.e. the sum of technology costs and programme delivery costs, for all nine programmes. Of course, only testing of these programmes under local conditions can provide the data to reconfirm this forecast.

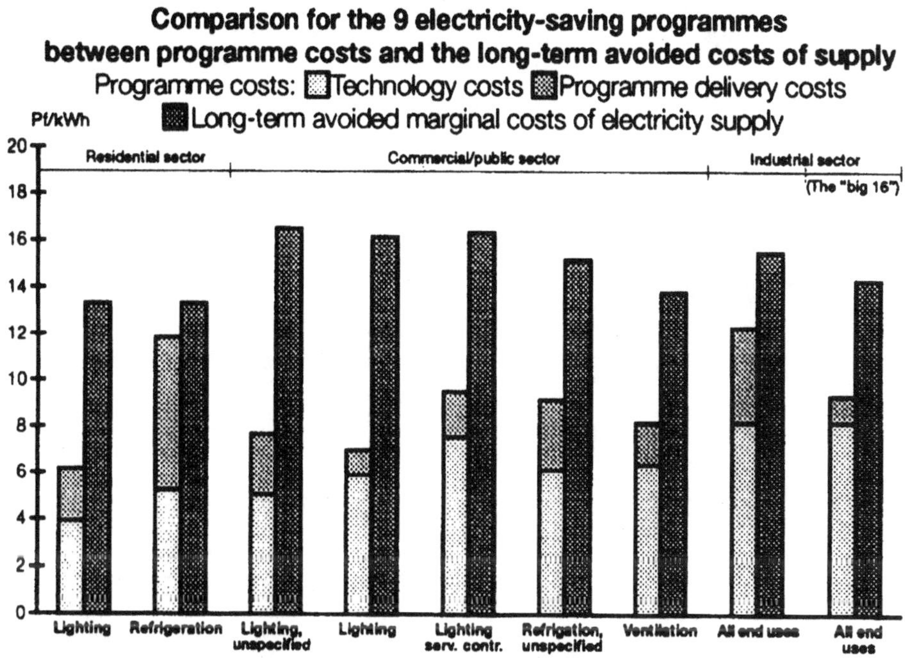

Figure 4.
Costs and benefits from nine DSM programmes for Stadtwerke Hannover.

Fig. 4 also displays the costs for the utility to deliver the programmes, including programme design, marketing, administration, and evaluation costs. Incentive payments to customers are not included here since they are not real costs to society. They are only transfer payments from one part of society (the utility) to another (the participants of the programmes).

5. IRP and economic incentives

By implementing these nine programmes, Stadtwerke Hannover can reduce consumption by 200 million kWh per year and achieve a net benefit to their customers of about 5,5 million ECU (about 10 million DM) per year. These higher incomes, in turn, may generate higher consumption of all kinds of goods. Higher consumption will lead to better employment, which is also spurred directly by higher investment in energy-efficient equipment due to the programmes.

Simultaneously, emissions will be reduced, and an additional benefit from avoided externality costs will be created. This externality benefit is calculated to about 3,5 million ECU (about 6 million DM) per year using the data from the GEMIS model (Öko-Institute / GhK 1995).

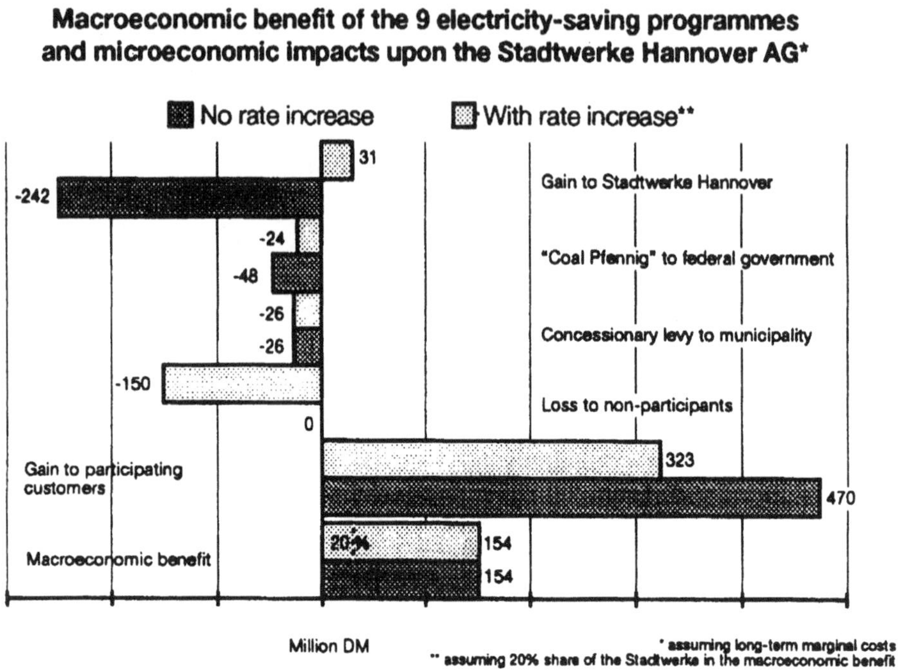

Figure 5.
Economic perspectives on nine DSM programmes for Stadtwerke Hannover without and with rate increase.

However, if electricity rates and prices remain constant relative to the situation without the nine DSM programmes, these net benefits will be distributed rather unevenly between the utility and its customers (Fig. 5). The dark columns in Fig. 5 show that there is a cumulated net macroeconomic benefit over the useful life of the technologies of 154 million DM, but Stadtwerke Hannover will loose 242 million DM over this 15 year-period. The reason for this is lost revenues, because energy savings due to the nine DSM programmes are reduced kWh sales for the utility. The participating customers are the big winners, gaining 470 million DM from rebates plus bill reductions due to reduced electricity consumption. Another loser is the municipality of Hannover since the concessionary levy which has to be paid by utilities in Germany is tied to kWh sales. This presents a disincentive to DSM programmes which should be removed by the Federal Government. The "Coal Pfennig", the national coal subsidy scheme, has been abolished with the beginning of 1996 (coal subsidies are now paid for by tax-payers, no longer by electric rate-payers).

But at constant prices, the disincentive to the utility due to the loss of 242 million DM is very high. Therefore, the utility should be allowed to raise its rates in order to recover DSM programmes costs just like costs for new power plants or transmission lines. Furthermore, the utility should be compensated for the difference between lost revenues and avoided marginal supply costs. And third, a moderate incentive for the utility to prefer DSM over expansion of supply is justified by the macroeconomic benefit of DSM. The regulatory authorities in Lower Saxony, which are responsible for regulating Stadtwerke Hannover, have said they will act in favour of such price increases if they are justified by cost-effective programmes.

Assuming a rate increase that compensates the net loss of 242 million DM and allows the utility to earn an additional 31 million DM, i.e. 20 % of the net macroeconomic benefit, completely turns the picture. The utility now has a gain of 31 million DM (white columns in Fig. 5). Customers participating in the DSM programmes still have a net benefit of 323 million DM. Two groups of losers remain: The municipality looses 26 million DM of concessionary levy, but gains from increased employment and maybe from the additional profit of Stadtwerke Hannover (31 million DM). Customers which do not participate in the DSM programmes loose 150 million DM.

Therefore, a specific feature of most cost-effective DSM programmes aiming at strategic reduction of electricity demand (kWh), not only capacity demand (kW), is that they lower total resource costs but raise rates. This is because costs decrease but sales decrease, too. Therefore, lower costs have to spread over fewer kWh, which normally leads to an increase in the price per kWh because costs decrease less than sales.

For the Hannover example, the examined price increases vary between 0,003 and 0,011 ECU/kWh (0,5 to 2,0 Pf/kWh) after ten years, compared to a situation without DSM. These increases are in most cases below the price decrease which occurred in 1996 in Germany due to the end of the national coal subsidy scheme.

Still, a rate increase due to DSM programmes causes a distributional effect between programme participants and non-participants. This should not be an argument against implementing DSM programmes. Rather, they should be designed in a way that allows all customers to participate in one part of a programme or another, e.g. through addressing a variety of technologies and end-uses, carrying on programmes long enough for every customer to have the opportunity to participate when a new appliance has to be bought anyway, and offering programmes for all customer groups.

It should be noted, however, that any costs for expanding electricity supply can also generate distributional effects between those customer groups which caused the expansion and those which did not or less than others.

However, it should be kept in mind that even after a possible rate increase, the bill of all customers for the energy services they require will be reduced by more than 80 million ECU (150 million DM) spread over the next 15 years in the Hanover case. Therefore, *the measure of economic efficiency changes with the introduction of IRP: it is no longer the price per kWh, but the total costs for energy services.* For the first time, the fact that there is competition between efficiency improvements on the demand side and the supply side is recognised correctly by IRP. IRP, in fact, aim to reduce existing market barriers for this competition between energy on one side, and know-how and capital for more energy efficiency on the other.

6. Is IRP applicable to the Central and Eastern European countries?

In principle, IRP/LCP is applicable to Central and Eastern European countries as well as to Western European countries. In practice, applicability depends especially on two features of the national situation:

First, the amount of electricity conservation potential which is cost-effective to society depends on two circumstances: is there a national primary energy production, or does primary energy and/or electricity have to be imported at world market prices? And are there national producers and providers of energy-efficient end-use technologies (e.g., CFLs) and services (e.g., installing energy-efficient ventilation/air conditioning systems), or do these technologies have to be imported?

The following table gives an indication for the prospects of electricity conservation at different combinations of these factors.

TABLE 3. Prospects of electricity conservation depending on the relation between demand-side and supply-side costs.

	domestic energy	imported energy
domestic efficiency products	depends	good
imported efficiency products	poor	depends

Second, the feasibility of IRP-oriented NEGAWatt programmes for utilities depends on the regulatory framework just the same as for Western European utilities. However, where there still may exist subsidised electricity rates for certain classes of customers, the avoided costs through NEGAWatt programmes may be higher than the lost revenue.

If this is the case, there is an incentive for the utilities to invest in demand-side energy efficiency even without the possibility of a rate increase.

7. A favourable legislative framework for IRP/LCP

A favourable framework for IRP/LCP should foster competition between energy and capital know-how, not just between energy suppliers and energy carriers for the lowest prices. Therefore, unbundling of distribution and supply and introduction of retail competition between suppliers for retail customers will prevent utilities from implementing large-scale NEGAWatt programmes (Wuppertal Institute 1996).

The following table gives an assessment of the prospects of IRP/LCP for different combinations of unbundling between generation, transport and distribution on one hand and third-party access (TPA; either for distribution/supply companies, which remain regulated monopolies, or for customers, which creates retail competition through unbundling of distribution and supply) on the other. In all cases except the retail wheeling situation the public control authorities would require the compliance with IRP principles and would give corresponding incentives to distribution/supply companies.

TABLE 4. Prospects for integrated resource planning in the electricity sector for different combinations of unbundling and TPA concepts.

Degree of Unbundling between generation, transport and distribution / **Degree of Third Party Access**	**by ownership**	**by management**	**by accounting**
for distribution/supply companies (wholesale competition)	++	+	depends
for distribution/supply and large industrial companies	+	+	depends
for all customers (retail competition)	-	--	--

Therefore, key elements of a regulatory framework for the introduction of IRP/LCP are:

1. Structural elements of the electricity system:
- closed service territories for all but the largest customers enable quality competition;
- integration of distribution and supply is necessary to finance conservation power plants through rates;
- unbundling of generation and distribution/supply is helpful, because it removes the incentive to the distribution/supply company to increase sales in order to use its generation facilities fully;
- tendering procedures for new capacity are more favourable than licensing procedures, since a precondition for a tendering procedure is forecasting the need for and controlling the construction of new capacity.

2. Tariff and pricing policy
- incentive regulation to make NEGAWatt profitable for utilities:
- approval of costs for cost-effective NEGAWatt programmes;
- key element is the decoupling of sales and profits to remove the disincentive to the utility to reduce kWh sales;
- a positive incentive to make NEGAWatts even more profitable for the utility than MEGAWatts.

Recently, the European Commission proposed a directive for the introduction of IRP in the electric and gas distribution sectors of the EU member states. It was approved by the European Parliament with great majority on first lecture. This directive would provide that Member States shall:

- ensure that electricity and gas distribution companies file IRP plans regularly;
- examine if cost-effective DSM options are realised;
- ensure that utilities can cover DSM expenses as well as lost fixed cost contributions;
- encourage utilities to implement
- comprehensive information programmes;
- rebate programmes;
- energy conservation programmes for low-income customers;
- third-party financing activities;
- stimulate the inclusion of DSM into tendering procedures for new capacity.

This directive is necessary to introduce IRP on a broad basis in the EU. It must be seen *as an integral part of a well-understood internal energy market* that emphasises the competition between energy on one hand and capital plus knowledge for demand-side energy efficiency on the other.

8. Conclusions

A bright future for IRP and DSM in Europe is justified by the emission reductions and the macroeconomic benefits that can be achieved. The possibility of rate-financed energy-efficiency programmes is open especially to grid-connected energies like electricity and gas, where there exist natural monopolies for the transport and distribution sectors. This is the main reason why the proposed IRP directive is correctly concentrating on the electricity and gas sectors, and why it is necessary to have an EU directive for introduction of IRP.

The directive is necessary to harmonise the framework for distribution/supply utilities in all EU countries to generate a competition for quality of energy services instead of a price competition for kWh. Only if energy efficiency programmes are offered to the customers of all distribution/supply utilities across Europe, it will be possible to avoid economic inefficiencies. Without such harmonisation, utilities offering rate-financed energy-efficiency programmes would have to fear price competition from utilities not offering such programmes. This would lead to a situation where no utility would want to perform such programmes.

Therefore, another favourable condition for the implementation of IRP is *not* to introduce retail competition between suppliers for final customers, except for very large customers. Instead, the functions of distribution and supply should remain integrated with distribution/supply utilities. These should have closed service territories, but should be regulated according to IRP principles: giving them an incentive to sell less kWh whenever this reduces total costs to their customers.

9. References

1. ELSAM (eds.) (1994) *Integrated Resource Planning: From Concept to Practice, Integrated Resource Planning in the Danish Electric Utilities, Main Report,* ELSAM, Fredericia.

2. Elser, M. (1994) Innovative Energiedienstleistungskonzepte für Industrie und Gewerbe in Dänemark, Norwegen und Schweden, Freiburg.

3. Goldman, C.A. and Kito, M. F. (1994) *Review of Demand-Side Bidding Programs: Impacts, Costs, and Cost-Effectiveness*, Lawrence Berkeley National Laboratory, Berkeley.

4. Mills, E. (1991) *Evaluation of European lighting programmes, Utilities finance energy efficiency*, Energy Policy, April 1991.

5. Öko-Institute / GhK (Gesamthochschule Kassel) (1995) *Umweltanalyse integrierter Energie-, Transport- und Stoffsysteme: Gesamt-Emissions-Modell Integrierter Systeme (GEMIS) Version 2.1, Aktualisierter und erweiterter Endbericht* im Auftrag des Hessischen Ministeriums für Umwelt, Energie und Bundesangelegenheiten (HMUEB), U. R. Fritsche u.a., Darmstadt/ Freiburg/ Berlin/ Kassel (published by HMUEB, Wiesbaden).

6. Öko-Institute / Wuppertal Institute (1995) *Integrated Resource Planning, The Hanover LCP Case Study, Final Report*, Stadtwerke Hannover, Hannover.

7. Thomas, S. (1995) Evaluation of RWE's KesS Rebate Programme for Efficient Residential Appliances, in A. Persson (ed.), *Proceedings of the 1995 ECEEE Conference*, ECEEE, Stockholm.

8. VDEW (Vereinigung Deutscher Elektrizitätswerke) (1994) DSM-Projekte Deutscher EVU, Stand Mai 1994, Frankfurt/M.

9. Wuppertal Institute / BEM (Beratungsgruppe Energie+Marketing) (1995) *Evaluierung des KesS-Programms der RWE Energie AG*, Peter Hennicke u. a., Wuppertal/Icking; RWE Energie AG, Essen.

10. Wuppertal Institute (1996) *Rational Planning Techniques (RPT): A Tool to Enhance Energy Efficiency?* Report to the European Parliament, Wuppertal; European Parliament, Brussels.

APPLIANCE ENERGY LABELLING AND EFFICIENCY STANDARDS IN EUROPE

PAUL WAIDE
PW Consulting
5 Oak Road
Cheshire M33 2FD, UK

1. Abstract

This paper summarises electric appliance usage, efficiencies, and technical saving potentials in Europe and examines the role of minimum energy efficiency standards as a policy mechanism to improve appliance energy performance. The main geographical focus is on the states of the European Union but details of programmes in Switzerland and Eastern Europe are also presented. This paper is an updated version of a paper 'Appliance Energy Standards in Europe' which was published in *Energy and Buildings*, Vol 26 (1), in 1997.

2. Introduction

In recent years European governments have introduced a range of policy measures aimed at improving the energy efficiency of domestic and commercial appliances sold in their borders. These measures have been in response to a recognition by many experts and policy makers that market competition will not on its own produce the level and speed of improvement in appliance energy efficiency that is considered to be achievable and desirable. The principal motivation for these measures has been the wish to fulfil macro-environmental goals such as greenhouse gas emissions reduction targets most recently committed to at the Kyoto conference of the United Nations Forum for the Climate Change Convention but also national energy policy objectives such as the nuclear power moratoriums in Sweden and Switzerland. Consumer cost-benefit arguments that have traditionally had a more central role in North America have also been an important policy aim as has the long standing desire to improve energy security.

S. Nørgård et al. (eds.), Development with Sustainable Use of Electricity, 237-286.

In particular though, under the terms of the Kyoto agreement the European Union is committed to reduce its annual greenhouse gas, and hence CO_2 emissions to 8% less than 1990 levels by 2008-12. Yet it is difficult to envisage how these targets can be met should total electricity demand in the EU continue to grow at more than 2% per annum. Certainly, the options to reduce emissions from generation through the substitution of high CO_2 emitting fuels by low emitting fuels can only be pursued so far without incurring significant additional supply side costs. By contrast there remain a wide range of options to raise the energy efficiency of electricity using equipment that offer overall cost reductions for consumers. These so called 'win-win' options are the primary focus of policy makers eager to balance their CO_2 emission budgets.

Historically, European states were among the first to introduce legislation to limit the energy consumption of domestic appliances during the nineteen sixties and seventies[1]. However, much of this was weak and poorly implemented and thus had little impact on actual appliance energy consumption. As a result of pressure to harmonise European trading conditions most of this national legislation was repealed in the late nineteen seventies and early eighties. Yet by the early 1990s growing concern about climate change and a realisation that the national policy vacuum needed to be filled from the centre led to the European Commission[2] (the executive arm of the EU) being given responsibility to coordinate EU efforts to improve appliance efficiency. Under the auspices of the SAVE and SAVE-II programmes[3] of the Directorate General for Energy, a spectrum of policy measures are being enacted including:

- commissioning of several major technical studies;
- development of energy and performance testing protocols;
- introduction of legislation for mandatory energy labels across a range of appliances;
- establishment of voluntary energy efficiency targets with manufacturers;
- introduction of mandatory minimum energy efficiency standards, MEES.

In some cases these measures have been complemented at the national level by:

- instigation of technology procurement programmes to promote more rapid development of high efficiency appliances;

[1] France introduced mandatory minimum energy efficiency standards for refrigerators in 1966 [1] and for freezers in 1978 [2]. Russia introduced mandatory energy efficiency standards in 1976 [3], and Poland is believed to have had mandatory minimum energy efficiency standards for a range of electrical appliances from as early as 1962 [4].

[2] Hereafter called the 'Commission'.

[3] SAVE was the principal energy efficiency programme of the EU adopted on 29 October 1991 which ran until 1995. SAVE-2 has superseded SAVE and will run from 1996 to 2000. The SAVE programme includes initiatives in all energy-consuming areas of the economy (homes, buildings, transport, industry, etc.) but the promotion of energy efficient appliances is one of the main focus areas of the programme.

- instigation of purchasing procurement programmes to ensure a mass market for more efficient products;
- establishment of award labels for the most environmentally benign products;
- issuing of appliance product lists indicating appliance energy consumption; and
- enactment of utility or government funded rebate schemes for efficient appliances.

In addition, the Directorate General for the Environment has complemented the energy labelling scheme by introducing a voluntary award label for the most environmentally benign products that uses energy efficiency as one of the most important criterions.

A listing of European policy actions by appliance type is provided in Appendix A.

Under the SAVE programme the Commission has currently identified four broad appliance groups where it intends to act to improve their efficiency. The groups, with the Commission's estimate of their current consumption and the Commission's saving targets, are indicated in Table 1.

TABLE 1. The SAVE Programmes EU Appliance Energy Use Estimates and Savings Targets [5]

Appliance	**Total Consumption (TWh/yr)**	**Potential Savings %**	**Potential Savings (TWh/yr)**
Cold Appliances[4]	120	30	40
Clothes-washers	35	25	9
Water heaters	100	20	20
Clothes-dryers	10	10	1
Dishwashers	15	30	5
TVs and VCRs (standby	13	80	10
consumption)	340	12	40
Industrial Electric Motors[5]	100	30	30
Office Equipment	150	20	30
Commercial Lighting[6]			
Total EU electricity consumption savings of 140 TWh e.g. 10% of all use.			

This paper presents an overview of European governmental programmes to improve appliance energy efficiency and, in particular, the role of minimum energy

[4] Cold appliances are appliances used for the cold storage of food in the home i.e. refrigerators, freezers and their combinations. The author estimates that the current EU consumption of these appliances is 119TWh/yr which corresponds to a value 11TWh/yr less than the Commission's figure.

[5] Defined as motors of power greater than 1 kW in the industrial sector. De Almeida [6] estimates that industrial motors of this type in the EU consume a greater amount ≈ 400TWh/yr.

[6] BRE [7] estimate that lighting in the EU service sector accounted for 111.37TWh in 1993.

efficiency standards and energy labelling. The main focus is on the residential sector where the greatest activity has occurred.

3. The need to reduce appliance electricity consumption

In the eighteen years between 1974 and 1994, electricity consumption in the EU states increased from 1052 TWh/yr to 1728 TWh/yr. Over the same period, total emissions of CO_2 from the EU states remained almost static at around 3120 million tonnes per year. Although the share of EU electricity produced by thermal generation fell from 79% in 1974 to 51% in 1994, the fraction of thermal electricity generated by coal increased from 47% in 1974 to 63% in 1994. For this reason and because only modest improvements have been attained in thermal generation efficiency, the average quantity of CO_2 released per kWh generated has only reduced slightly since 1974 and stood at 432 tonnes of CO_2 per GWh generated in 1994 [8]. Accordingly, the contribution of the electricity sector to total EU CO_2 emissions continues to increase almost as fast as the growth in electricity demand and, because the share of thermal generation is increasing again, it is projected to do so into the future [9].

The growth in electricity consumption is not evenly distributed among sectors. The fastest growth has occurred in the commercial and public sector, which averaged 4.0% per annum between 1980 and 1994. Over the same period, the growth in consumption for the residential sector averaged 2.5% p.a., and in the industrial sector, 1.4% [10, 8].

Breaking total EU electricity consumption down by sector, it is found that the residential sector uses approximately 31% of the total (537 TWh/yr for the EU); the commercial & public sector consumes about 25% (429 TWh/yr for the EU); and the industrial, transport & agricultural sector uses the rest (762 TWh/yr) (see Figure 1).

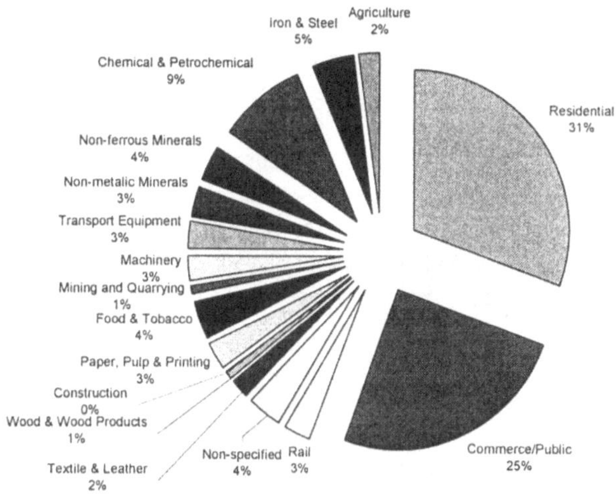

Figure 1.
Total electricity consumption in the EU, by sector = 1728 TWh/yr [8, 10]

In the EU commercial & public sector, it is estimated that 150 TWh/yr are used by motors of greater than 0.75 kW power [11], and about 111.4 TWh/yr[7] for lighting [7]. HVAC, which will include some of the previous motor consumption already mentioned, accounts for about 32% of electricity consumption in the service sector [7] In the EU industrial sector, motors of over 0.75 kW power are estimated to consume approximately 480 TWh/yr [11].

The total consumption in each sector is influenced by the number (or stock) of electrical appliances (which in the case of the residential sector is linked to the number of households and the ownership of appliances per household), the usage of the appliance, the size or capacity of the appliance, the features offered with the appliance, and the efficiency of the appliance. The level of service and energy demand for appliances is further influenced by external physical factors such as: climate, quality of building fabric insulation, thermal controls, and the magnitude of passive gains for electric space heating demand; availability of natural day lighting for lighting demand; the relationship between detergent wash performance and wash temperature for clothes washers; and kitchen temperature for cold appliances, etc.

[7] As of 1994 the EU commercial lighting consumed 93TWh/yr and EFTA 32TWh/yr.

3.1. ELECTRICITY DEMAND IN THE RESIDENTIAL SECTOR

3.1.1. Trends in Past Consumption

The ownership and use of domestic electric appliances is continuing to grow rapidly in Europe, although the rate of increase appears to be slowing. It is convenient to separate out trends in residential electricity consumption by appliances that are electricity specific (i.e. those for which electricity is the only practical fuel), and trends among appliances for which alternative fuels are readily available. Gas is the main competitor of electricity for the second class of appliances, especially for water heating, space heating[8] and cooking, and the electricity usage for these end-uses varies considerably from country to country within Europe depending on the availability of cheap domestic gas. In the UK, for example, annual residential electricity use for space and water heating reached a peak of 47 TWh in 1972 before declining with the advent of cheap North Sea gas to 30 TWh in 1993 [12]. In France the situation was the reverse with electricity use for space and water heating increasing from 8.6 TWh in 1973 to 49.7 TWh in 1991 [13]. Annual space heating and cooling demand is also strongly influenced by variation in the annual average climatic conditions.

The trends in electricity specific appliance energy consumption are much more uniform across large geopolitical groupings. Analysis of the UK electricity market has shown that consumption by electricity by domestic lights and appliances grew from 39.3 TWh/yr in 1970 to 71 TWh/yr in 1994 [14]. Figures for France indicate that electricity consumption by electricity specific domestic appliances has grown from 19.7 TWh/yr in 1973 to 49.2 TWh/yr in 1991. The UK analysis further identifies that the energy consumption of electricity specific domestic appliances has been the fastest growing source of carbon dioxide emissions outside the transport sector.

No comprehensive study has yet been done to aggregate all residential end-uses in Europe in order to ascertain their individual contributions to total electricity demand although in some cases this has been done at the national level. Figure 2 shows one estimate of the evolution in residential electricity consumption by 5 major end-uses for the four largest countries in the EU [15].

As with the more detailed UK data, Figure 2 clearly illustrates the comparative importance of the growth in electricity specific appliance consumption within the residential sector.

---

[8] Oil is also a popular fuel for space heating in some parts of Europe.

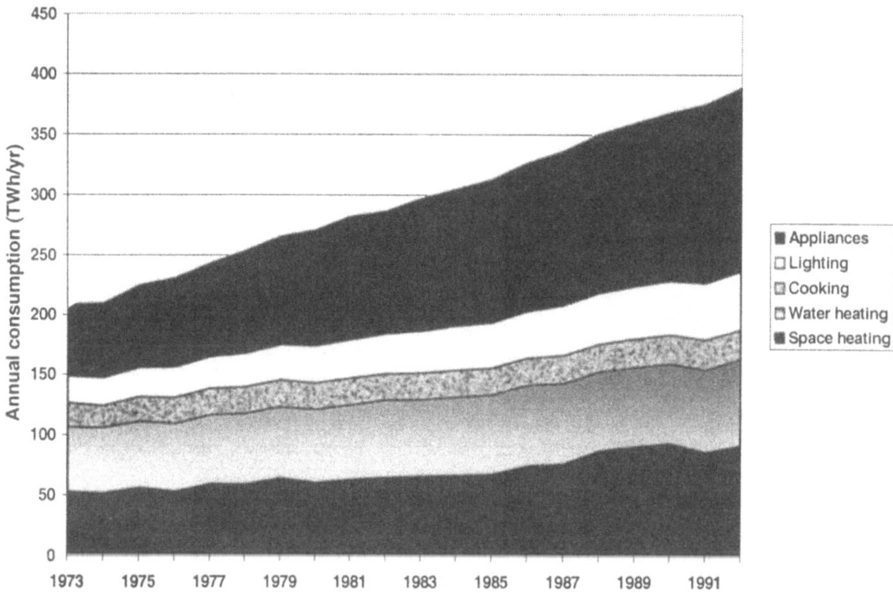

Figure 2.
Estimated Residential Electricity Consumption by End-Use in the EC-4 (UK, France, Italy, Germany). [15]

3.1.2. Current Situation

Within the EU residential sector, electricity specific usage is dominated by a few end-uses, namely: food refrigeration ≈109 TWh/yr (20.3%)[16], lighting ≈74 TWh/yr (13.8%)[9], home laundry ≈44 TWh/yr (8.2%)[17], brown goods (e.g. TVs, VCRs, etc.) ≈40.8 TWh/yr (7.6%)[18], cooking ≈32 TWh/yr (6.0%)[9] (of which ≈ 9.2 TWh/yr may be for microwaves, [19]), small and other appliances (excluding cooking) ≈ 54 TWh/yr (10.1%)[9], and lastly dishwashers ≈16 TWh/yr (3.0%)[17], see Figure 3.

[9] Based on scaling up the average household consumption from UK, France, Italy, Netherlands, Sweden, Portugal, Austria, Denmark and Finland (60% of EU households) to the whole EU.

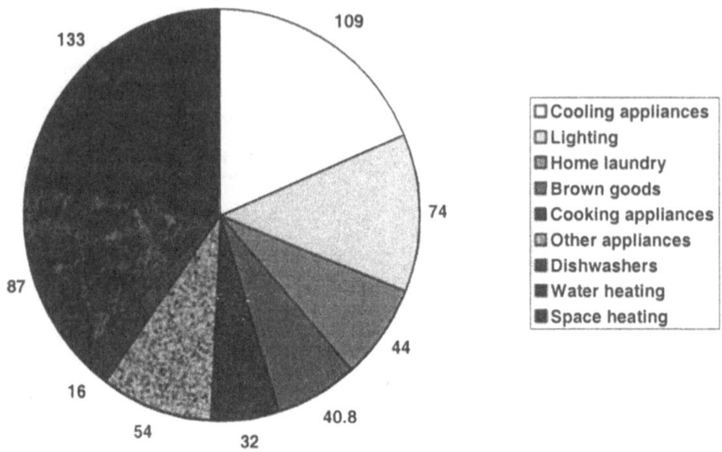

Figure 3.
Estimate of Residential Electricity Consumption by End-Use within the European Union (TWh/yr).

Two of the largest uses of residential electricity are not electricity specific; namely, water heating, with an estimated ≈87 TWh/yr (16.2%)[20], and space heating, with an estimated ≈133 TWh/yr (24.8%)[9]. The fact that the quoted percentage shares give a total of 109.8% is an indication of the uncertainty that exists in the figures. All the values are estimates, and many of them are calculated by scaling-up estimated household consumption data for nine EU countries to the level of the whole EU. The cold and wet appliance consumption data is the most accurate, but there is considerable uncertainty concerning the two major end-uses of lighting and space heating. Air conditioning is still quite uncommon in European homes and therefore is not a significant residential end-use at the present time.

It should be emphasised that although the consumption of electricity specific domestic appliances is increasing in all European states, there are very significant differences in household electricity consumption between European nations by end-use.

3.1.3. European Cold Appliance Efficiency Trends

Cold appliance energy efficiency improved quite appreciably between 1970 and 1985 partly as a response to the market driven introduction of better technology and partly as a result of various government policy interventions; most notably a voluntary agreement between industry and government in Germany that operated from 1980 to 1985. The average energy consumption of new refrigerator-freezers and freezers fell by from 40-50% over this period, even though their size generally increased. However, from 1986 to 1993 the energy efficiency of new cold appliances showed a slight deterioration as manufacturers reduced the amount of CFCs they were using in foam insulation and searched for a replacement refrigerants in response to the demands of the Montreal Protocol on protection of the ozone layer. In 1992 it became clear that the EU was poised to introduce mandatory energy labelling for cold appliances, which were possibly to be followed by minimum energy efficiency standards, and this dual policy development seems to have stimulated manufacturers to concentrate on designing more efficient products. In Germany the efficiency of the market improved by 15.5% between 1992 and 1995; in France the efficiency of the market improved by 16.1% from 1993 to 1996; in the UK the efficiency of refrigerator-freezers increased by 7.3% from 1994 to 1996; while in the Netherlands the efficiency of the market improved by 12.6% from 1992 to 1995, see Figure 4.

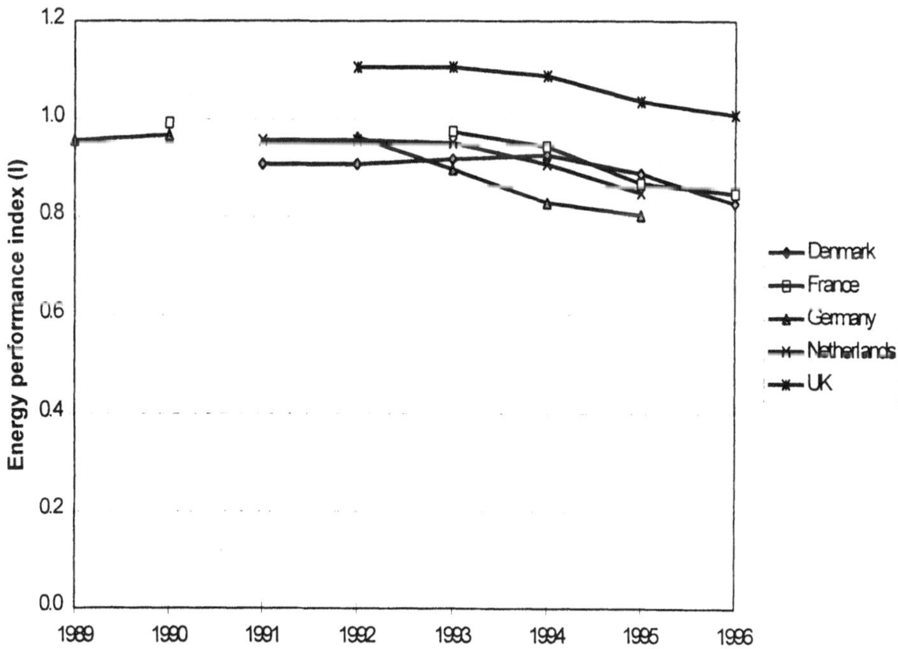

Figure 4.
Evolution of refrigerator-freezer energy efficiency in five EU countries [16]

3.1.4. Efficiency Trends for Other Residential Appliances

The energy efficiency of other appliances has also improved over time, although not always as consistently as for the cold appliances. The energy used by the average European washing machine, based on the now out-dated IEC 456 test, has declined from about 3.25 kWh/wash in 1973 to about 2.05 kWh/wash in 1994. Over the same period the energy efficiency of the entire clothes washing and drying process has increased significantly largely because of lower wash temperatures made possible by more efficient detergents and higher spin speeds. For dishwashers, the average consumption has declined from about 2.75 kWh/load in 1975 to 1.60 kWh/load in 1994 [17]. With both clothes- and dish-washers, the rate of market driven efficiency improvement is forecast to slow in the future [17]. The little historical data available for clothes dryers indicates no clear energy efficiency improvement between 1987 and 1993 [17]. The average energy intensity of electric ovens in Denmark is estimated to have fallen by 13% between 1970 and 1990 [21], while in Austria the average improvement in oven efficiency from 1977 to 1990 was closer to 30% [22]. The average power demand for new TVs in Austria operating in the on-mode fell from around 225W in 1977 to just over 50W in 1990. The average efficiency improvement of electric water heaters with greater than fifteen litres of storage capacity on the Austrian market between 1977 and 1990 was 60%. For electric water heaters with less than 15 litres of storage capacity, it was 55% [22].

4. The European Appliance Industry and Market

Approximately 56 million large appliances were purchased in Western Europe in 1992, and an additional 17 million in Eastern Europe out of a world total of 194 million [19]. These figures make Europe the world's largest regional market for white goods[10] with an annual value of DM45 billion for Western Europe alone in 1994 [23]. This figure (about 40% of the global market and about 1.6 times the value of the US white goods market) is explained in part by the relatively high cost of white goods in Europe, that are among the most expensive in the world; however, this large market also reflects the consuming habits of a large and comparatively wealthy population.

Not surprisingly, appliance ownership levels are generally high, as indicated in Figure 5. However, the notion of Europe as a single homogenous market is very misleading since there are substantial differences in consumer preference and behaviour from country to country. For example, in 1994, 77% of households were using at least one freezer in Switzerland but only 18% in Spain and merely 1% in Turkey. National differences in appliance ownership patterns can be partially explained by differences in per capita wealth; however, cultural and geophysical factors are often equally or more

[10] Defined as: refrigeration appliances, home laundry appliances, large cooking appliances (including microwave ovens) and dishwashers.

important. For example, 39% of households possess an electric clothes dryer in England compared to 10% in Austria and only 1% in Italy.

These differences in consumer preference and behaviour can translate into significant differences in the aggregate appliance energy and power demand, e.g. in the UK, ownership of electric kettles is extremely high and it is common for evening peaks in power demand to occur because a large proportion of households activate their kettles simultaneously during the commercial breaks of popular television programmes. In other European countries, electric kettle ownership is usually much lower than in the UK and usage patterns less predictable.

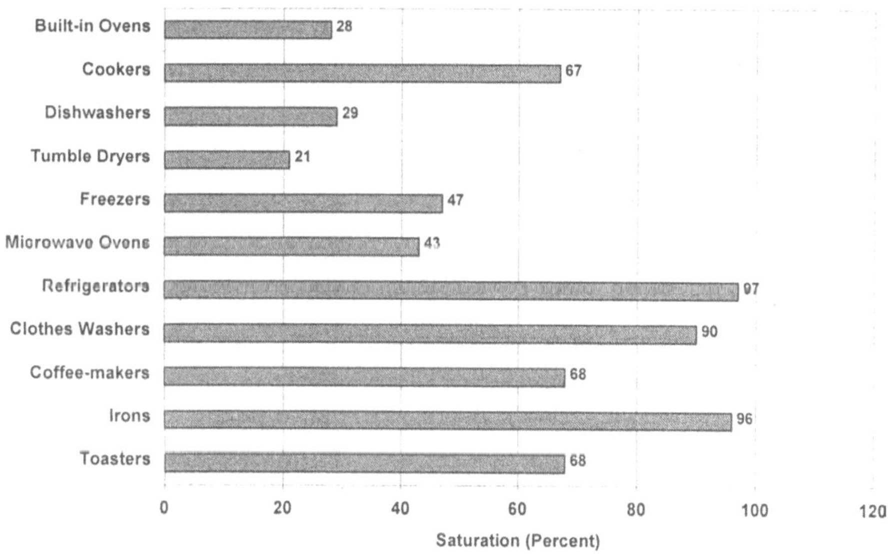

Figure 5.
Average Appliance Saturations in Europe for 1994 [23]

The European domestic appliance market is relatively fragmented compared to the North American or Japanese markets. There are in-excess of forty white-goods manufacturers with well over a hundred brands of which the top five manufacturers have 62.9% of market share; however, the trend has been toward rationalisation through acquisition. In 1970, about 400 companies served 80% of the market but by 1993 this figure had reduced to around 10-15 manufacturers [24]. Six of the world's top ten appliance manufacturers (ranked by value of sales) were European in 1992, including the largest, Electrolux.

Of the total 1992 West European white goods market, wet appliances (home laundry and dishwashers) accounted for 36.9%, cold appliances (refrigerators, freezers, and their combinations) accounted for 30.3%, and cooking appliances were 21.9% by value [24].

A snapshot of sales volumes by residential appliance type for each European country is given in Table 2.

TABLE 2. Appliance Sales in Europe (1993) [25]

Country	Cold Appliances	Cookers	Microwave Ovens	Washing Machines	Tumble Dryers	Dish-washers
Belgium	375	116	230	243	150	102
Denmark	411	105	75	124	55	73
France	2639	2292	1163	1982	546	712
Germany	3223	3053	1877	2615	657	1128
Greece	290	173	27	200	19	35
Ireland	94	76	42	143	17	12
Italy	2458	2042	665	1181	4	517
Luxembourg	16	5	10	10	6	4
Netherlands	570	334	385	440	240	79
Portugal	1077	356	52	244	15	88
Spain	1357	1533	625	1122	125	301
UK	1989	1559	1135	1470	520	299
Austria	756	352	171	234	34	139
Finland	205	120	98	119	13	54
Norway	207	80	78	116	35	61
Sweden	465	200	235	165	65	116
Switzerland	358	139	73	119	46	116
Bulgaria	86 [a]	95		92	192	485
Czech/Slovak Rep	1147	282		227		
Hungary	606 [a]	127		140	113	
Poland	813 [a]	322		790		
Romania	221 [a]	350		298		
USSR (former)	4794			6130		
Yugoslavia (former)	572	248		284		

[a] Not all types of cold appliances are included in this figure

There are substantial differences of interest between European nations. Nine of the worlds top ten white goods importing nations are in the European Union (the US is the fifth biggest importer) and seven out of the worlds top ten white goods exporting nations are European. However, most of this trade is internal to Europe with Germany and Italy dominating exports and France, Germany, and the UK dominating imports. Appliance retail also varies substantially by country, especially the relative dominance of the large specialist electrical retailers, discount stores, hypermarkets and small independent retailers.

5. Appliance Energy Testing Protocols and Certification in Europe

Before any policy that aims to influence the energy efficiency of new appliances can be introduced it is necessary to have a reliable, routine and repeatable energy or energy efficiency measurement. In most of the world product testing and certification agencies have been developed that issue test protocols describing how measurements should be made, establish quality control guidelines and correct practice requirements for testing laboratories and establish a regulatory framework under which product testing must occur.

The establishment of the test standards agencies CEN and CENELEC was an important component of the market harmonisation process embodied in the original Treaty establishing the European Economic Community and later in the Single European Act of 1986. These bodies now issue European appliance testing protocols and safety standards which apply in all the nations comprising the European Economic Area (EEA) (i.e. the EFTA and EU states). Some Eastern European countries also use the CEN and CENELEC refrigerator testing protocols and Japan has recently been granted permission to use the EN 153 protocol for cold appliances. Neither CEN nor CENELEC standards are mandatory within any European state unless they are incorporated in separate government legislation; however, in practice CEN and CENELEC standards are almost always adopted by the standards bodies within all these states. There is an opt-out clause within the European Treaty which stipulates that local trade protocols within EU member states can take precedence over the European protocols in some circumstances although the legal precedents for doing so are complex. In order to ensure the uniform implementation of standards within Europe, the European Organisation of Testing and Certification (EOTC) was established in 1993. The EOTC has formed a group to cover electrical products and to ensure that European testing laboratories and certification bodies recognise each other's work. In general, however, all the European energy labelling and MEES Directives rely on self-declaration by manufacturers.

The existence of accepted European energy and performance protocols by appliance type is indicated in Appendix A.

6. Minimum Energy Efficiency Standards

Many appliance efficiency analysts believe that mandatory minimum energy efficiency standards (MEES) are often the most effective means of rapidly improving the energy efficiency of appliances that have a mature market, high economic but currently unrealised technical energy efficiency potential, and a relatively slow rate of product evolution. If designed properly they can provide significant longer-term benefits to the consumer and the environment without penalising manufacturers.

Compared to alternative policy initiatives, MEES have the advantage that their benefits are comparatively simple to quantify, they treat all manufacturers, distributors and retailers equally, and the resulting energy savings are comparatively guaranteed. The impact of MEES legislation varies considerably depending on whether they are 'under-pinning' i.e. that they only prohibit the least efficient models from the market, or 'market-forcing' i.e. that they oblige manufacturers to develop a new generation of appliances with higher energy efficiencies than the current norm. In the latter case the energy savings from mandatory standards can be appreciable.

Voluntary energy efficiency targets negotiated between government and manufacturers are an alternative option to MEES. They have the merit of being less controversial and hence sometimes, but not always, easier to enact. Yet they can disadvantage signatory manufacturers in comparison with non-signatory manufacturers (typically domestic manufacturers as opposed to importers). The fact that voluntary standards are seldom universal and are not obligatory means that it is very difficult to persuade manufacturers to agree to a measure which will either reduce their profit margin or will increase the purchase price of their goods. This arises because of the legitimate fear that competitors will not follow suit. For MEES, a small increase of production cost is usually anticipated, but because in principle it applies to all manufacturers equally, it can be safely passed on to the consumer without losing market share. If the MEES are designed properly, the consumer will more than recoup the increased purchase cost through reduced operating costs, while manufacturer sales and profits should not be adversely effected.

6.1. THE STATUS OF APPLIANCE EFFICIENCY STANDARDS IN THE EU

6.1.1. MEES legislation in the European Union

Despite the potential for large energy savings, some EU member states have been unwilling to embrace mandatory energy efficiency standards, chiefly out of concern that they might adversely impact their appliance industry. In contrast some other member states have strongly advocated their introduction.

In the past this has resulted in an impasse that has left the European states still lagging behind North America when it comes to the enactment of effective legislation to improve the energy efficiency of appliances. In the EU, there is no framework legislation establishing the right of the Commission or another competent body to introduce MEES for selected products. Instead, for mandatory minimum energy efficiency standards to be passed, it is necessary to seek separate approval on an appliance by appliance basis from the Council of Ministers and the European Parliament. Decisions in the European Parliament are taken by a simple majority vote, but in the Council of Ministers they are taken by qualified majority voting, which means that only a minority of votes[11] need be cast against new legislation for it to be blocked. The strong emphasis of the EU decision taking system toward consensus can make the passage of proactive legislation difficult and lengthy.

Until recently the lack of a Framework MEES Directive and the controversy that surrounded the introduction of the cold-appliance MEES Directive discouraged EU officials from seeking new legislation for other appliances. Instead, they have been focusing on the negotiation of voluntary energy efficiency agreements with

[11] The more populous states are awarded the most votes although the votes per capita ratio is higher for the less populous states.

manufacturers. However, there are now signs that when the Commission finds it is unable to negotiate a meaningful voluntary agreement it is prepared to propose MEES to the Council and Parliament. Voluntary agreements are under discussion or have been reached for: clothes washers, washer-dryers, TVs, VCRs, PCs, monitors, fax machines, copiers, printers, and industrial motors [26]. At present, many of these initiatives are quite new so it is not clear how many of them will lead to meaningful energy savings, although a precedent already exists in the case of Switzerland.

As of 4.3.98 minimum efficiency standards legislation had been enacted in Europe for domestic gas- and liquid-fuel fired boilers (92/42/EEC, date 21.5.92) and for domestic refrigerators, freezers and their combinations (96/57/EC, date 3.9.1996). The cold appliance minimum energy efficiency standards Directive is due to come into force on 18.9.1999, while the boiler Directive came into force on 31.12.1997. At the current time these are the only community-wide minimum energy efficiency standards applicable to domestic appliances; although maximum allowable standing losses for electric storage water heaters are regulated in some EU states through national building regulations. Further MEES Directives are possible for some appliances and the Commission are currently drafting MEES for lamp ballasts and are considering EU-wide MEES for electric storage water heaters.

The domestic boiler Directive was thought to be unlikely to have a significant impact on the market and is seen as an underpinning standard to prevent very poor performance models from being sold.

The cold-appliance Directive will have some impact on the performance of models sold in the EU, although the efficiency levels implied by the standard in this first Directive are far removed from the least life cycle cost efficiency levels. In this sense the standard is close to an underpinning standard but has some elements of the US-style market forcing standard. Nonetheless, the Directive anticipates that a second tougher standard level may be introduced sometime after 2001. The real test of this Directive will be the firmness of the efficiency levels set for the second round.

6.1.2. The EU Cold Appliance Efficiency Standards Directive

The cold appliance efficiency standards Directive classifies the products available on the market into almost the same fundamental categories used in the energy labelling Directive[12]. The standard is set in relation to a reference energy consumption line which corresponds to the average energy used by models available for sale on the

[12] The energy labelling Directive has 10 categories of cold-appliance while the standards Directive has only nine. The difference is that the 10th category in the energy label refers to multi-door appliances and any other product types which are not covered in the nine principal categories, while the standards Directive does not apply to these appliances, which in any case are rare on the European market.

EU market circa ~1991 as a function of their adjusted volume[13]. A separate reference line is calculated for each of the 10 product categories. The efficiency standard is defined as a ratio of the reference line for each category. It is based on a statistical analysis of the 1991 EU market and was originally estimated to improve the new model stock efficiency so that it would consume 15% less energy than if the standard were not applied, thus it became known as the '-15% standard'. However, the average efficiency of the market has improved since 1991 and by the time the standard is implemented in 1999 it will result in a lower energy saving than 15%. It has been estimated [27] that the first round of the current EU cold-appliance MEES Directive will save 120 TWh from 1999 to 2020 compared to a base-case scenario which assumes no fresh policy initiatives after implementation of the energy labelling Directive. This value equates to 4.9% of the forecast base-case cold-appliance consumption over the same period.

The formulas defining the minimum energy efficiency standard (actually a maximum energy consumption standard) and the reference line are given in Table 3 [16].

[13] This is the net volume (i.e. the effective storage capacity of the appliance) adjusted to take account of the different cooling service being offered by compartments which cool food to different temperatures. This idea, first applied in the USA, takes account of the fact that a unit of space in a freezer compartment, for example, is providing a higher cooling service than the same unit of space in a refrigerator from an energetic perspective, thus a corrective coefficient is applied so that all storage space can be treated on an equal basis.

TABLE 3. Product categories and energy efficiency standards for cold-appliances in the EU [16]

Product class			Maximum allowable energy use (kWh/yr)[1]	
Description	Category No.	Ω	MEES for year 1999[9,10]	Average performance reference line[11]
Freezers				
Chest Freezers	9	2.15	0.480AV+195 (E)	0.446AV+181
Upright Freezers	8	2.15	0.434AV+262 (E)	0.472AV+286
Refrigerators and Refrigerator-Freezers				
Refrigerator without FFC[2]	1	1.00	0.207AV+218 (D)	0.233AV+245
Refrigerator/Chiller[3]	2	0.75	0.207AV+218 (D)	0.233AV+245
Refrigerator with no-star[4] FFC	3	1.25	0.207AV+218 (D)	0.233AV+245
Refrigerator with 1-star FFC[5]	4	1.55	0.557AV+166 (D)	0.643AV+191
Refrigerator with 2-star FFC[6]	5	1.85	0.402AV+219 (D)	0.450AV+245
Refrigerator with 3-star FFC[7]	6	2.15	0.573AV+206 (D)	0.657AV+235
Refrigerator with 4-star FFC[8]	7	2.15	0.697AV+272 (E)	0.777AV+303
Multi-door refrigerators and other appliances	10	$=(25-T_C)/20$	0.207AV+218[12](D)	0.233AV+245[12]
			0.557AV+166[13](D)	0.643AV+191[13]
			0.402AV+219[14](D)	0.450AV+245[14]
			0.573AV+206[15](D)	0.657AV+235[15]
			0.697AV+272[16](E)	0.777AV+303[16]

The value given by the formula is to be compared against the measured rating under the test protocol EN153. AV is the adjusted volume given by: $AV = \sum V_c \times W_c \times F_c \times C_c$ as summed over all compartments in the appliance, and where:

- V_c = the net volume of a given type of compartment in the appliance,
- W_c is the weighting coefficient for that type of compartment (equal to 1 for the fresh-food (5°C) compartment and equal to Ω for the non-fresh-food compartment)
- F_c is a factor which equals 1.2 for no-frost compartments and 1 for other compartments.
- C_c is a factor = 1 for N and SN class (temperate and sub-temperate climate class appliances), X_c for ST (sub-tropical climate class) appliances, and Y_c for T (tropical climate class) appliances. X_c and Y_c are defined as follows:

	X_c	Y_c
Chiller compartment designed to operate at 5 °C < T_c < 12 °C	1.25	1.35
fresh food compartment designed to operate at 5 °C	1.20	1.30
0 °C compartment	1.15	1.25
no-star compartment designed to operate at -6 °C < T_c < 0 °C	1.15	1.25
1-star compartment designed to operate at T_c < -6 °C	1.12	1.20
2-star compartment designed to operate at T_c < -12 °C	1.08	1.15
3 and 4 star compartment designed to operate at T_c < -18 °C	1.05	1.10

[2] FFC = frozen food compartment, which refers to any compartment operating at less than –6°C. Normal refrigerator compartments are designed to operate at 5 °C.
[3] Refrigerator/Chillers are appliances designed to operate at internal temperatures of 5 °C < T_c < 12 °C.
[4] No-star compartments are designed to operate at -6 °C < T_c < 0 °C.
[5] 1-star compartments are designed to operate at T_c < -6 °C.
[6] 2-star compartments are designed to operate at T_c < -12 °C.
[7] 3-star compartments are designed to operate at T_c < -18 °C.
[8] 4-star compartments are designed to operate at T_c < -18 °C and to be able to freeze more than 4.5kG of food from 25°C to -18°C in less than 24 hours.
[9] The Commission of the European Communities Directive 96/57/EC.
[10] Letters in brackets are the associated grade of the energy label for models which exactly meet the standard e.g. the Commissions proposal for chest freezer standards would prohibit from sale some models in the E category and all those in the F & G categories of the energy label.
[11] The line defining the average energy consumption as a function of adjusted volume of models available on the EU market between 1990 and 1992 from the Group for Efficient Appliances , GEA,study [28].
[12] For multi-door models where the temperature of the coldest compartment is > - 6 °C.
[13] For multi-door models where the temperature of the coldest compartment is ≤ - 6 °C.
[14] For multi-door models where the temperature of the coldest compartment is ≤ - 12 °C.
[15] For multi-door models where the temperature of the coldest compartment is ≤ - 18 °C.
[16] For multi-door models where the temperature of the coldest compartment is ≤ - 18 °C and the compartment can freeze > 4.5 kg of food from 25 °C to - 18 °C in 24 hours.

6.1.3. The EU gas and liquid-fuel fired boiler MEES

In addition to cold appliances the only other appliance group, electric-specific or otherwise, which are currently subject to EU MEES requirements are boilers using gas or liquid fuels within the nominal rated power range of 4 kW to 400 kW. The MEES thresholds from Directive (92/42/EEC, date 21.5.92) are shown in Table 4 where Pn = the nominal rated power of the boiler (in kW). The efficiency thresholds are expressed in terms of 'useful efficiency' values which is the ratio of the calorific value of the energy transmitted to the water and the calorific value of the fuel, burned at constant pressure, used to heat the water. This ratio is expressed as a percentage for comparison with the threshold equations.

TABLE 4. Product categories and energy efficiency standards for gas and liquid-fuel fired boilers in the EU

Type of boiler	Power interval	Efficiency limits as a function of the nominal rated power		Efficiency limits under partial loads (30% of nominal capacity)	
	(kW)	Average water temperature in the boiler (°C)	Minimum efficiency thresholds (%)	Average water temperature in the boiler (°C)	Minimum efficiency thresholds (%)
Standard boilers	4-400	70	≥84+2*LogPn	≥50	≥80+3*LogPn
Low-temperature boilers(*)	4-400	70	≥87.5+1.5*LogPn	40	≥87.5+1.5*LogPn
Gas condensing boilers	4-400	70	≥91+1*LogPn	30(**)	≥97+1*LogPn

(*) Including liquid-fuel condensing boilers

(**) Water feed temperature entering the boiler

Although these MEES prevent highly inefficient models from being sold on the EU market, they are more market underpinning than market forcing and are therefore not thought likely to have a large impact on average boiler efficiencies.

6.1.4. The EU Electric Storage Water Heater Efficiency Standards Proposal

A recent SAVE study [20] that investigated the technical and policy options to improve the energy efficiency of Domestic Electric Storage Water Heaters (DESWHs) has recommended that MEES be introduced to curb water heater standing losses. It is estimated that DESWHs have standing losses totalling some 19 TWh/year across the EU and that these can be cost effectively reduced through the introduction of MEES.

The study has proposed that MEES efficiency levels are set so the standing losses $L_{st,st}$ (measured according to the test standard IEC 379, 3rd edition, 1987, in kWh/day) of a particular DESWH with rated capacity V (in litres) must not exceed $L_{st,mes}$, as shown in Table 5.

TABLE 5. Proposed minimum energy efficiency requirements as a function of rated capacity [20]

Type of DESWH	Capacity (litres)	Minimum energy efficiency requirements[14] ($L_{st,mes}$)
Vertical	> 50–1000	$0.2 + 0.051 * V^{2/3}$
Horizontal	> 50–300	$0.75 + 0.008 * V$
Small	5–50	$0.13 + 0.0553 * V^{2/3}$

The same study has recommended that energy labels be introduced in addition to MEES. The proposed standing losses efficiency thresholds for MEES and labels and their relation to DESWH models currently found on the EU market are shown in Figure 6.

[14] V = rated capacity in litres.

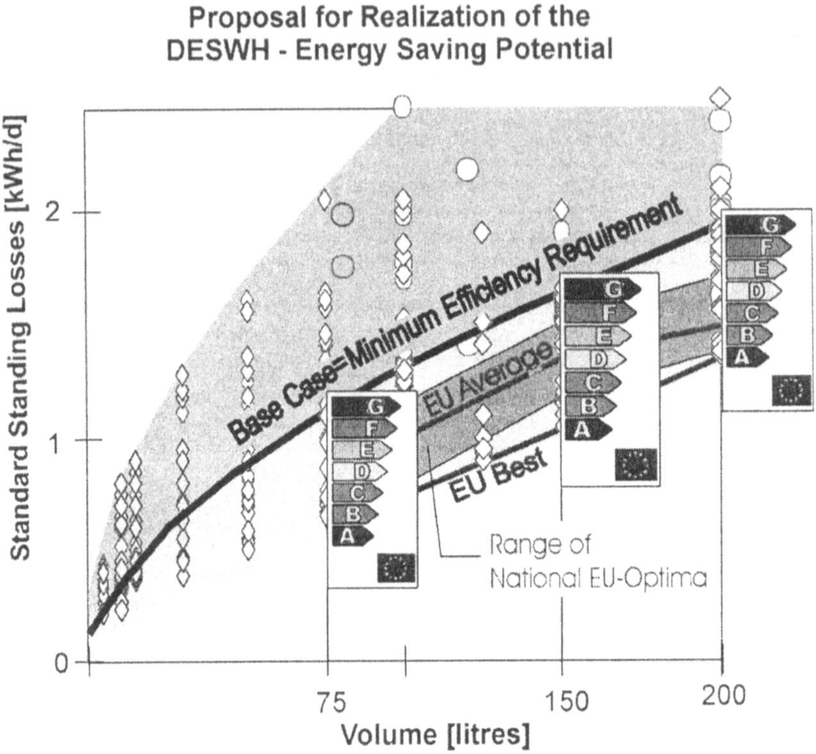

Figure 6.
DESWH standing losses for models on the EU market and the proposed new MEES and energy labelling efficiency thresholds. [20]

The forecast impact of the proposed MEES are shown in Table 6.

TABLE 6. Impact of the proposed DESWH minimum efficiency standards [20]

Annual impact	**2000**	**2005**	**2010**
Consumer cost savings (million ECU)	54	173	259
Change in manufacturers return on equity (%)	-0.4	0.0	0.0
Reduction in CO_2 emissions (Mt CO_2)	0.22	0.73	1.10
Reduction of electricity consumption (GWh)	452	1 445	2 158

As a result of this study the Commission has announced its intention to propose MEES legislation for DESWHs although they have given manufacturers until the end of June 1998 to propose an alternative voluntary agreement.

6.1.5. Minimum Efficiency Standards for Other Appliances?

New efforts are being directed toward other electric appliances. Studies of the energy use of office equipment [29], the electric lighting markets [7], the standby mode of TVs and VCRs [18], electric storage water heaters [20], and industrial motors [30] have been completed as a precursor for possible additional programmes in these areas. Work is underway on commercial refrigeration cabinets, the energy use of TVs in the on-mode, air-conditioners & heat pumps, and cooking appliances.

6.2. THE STATUS OF APPLIANCE EFFICIENCY STANDARDS IN EASTERN EUROPE AND RUSSIA

Currently, there are no mandatory minimum energy efficiency standards in place in the central Eastern European states, but MEES do exist in the Russian Federation and other CIS states. In September 1994, Russia signed a Memorandum of Co-operation with the US Department of Energy to initiate a new programme to promote appliance energy efficiency. This has the ambitious goal of harmonising Russian and US appliance standards, testing, and certification. The intention appears to be to bring Russian equipment up to the energy efficiency levels prescribed in the US NAECA legislation [31]. The appliances initially targeted are: refrigerators & freezers, electric motors, residential boilers, and lighting ballasts. Russia already has minimum energy efficiency standards for refrigerators, vacuum cleaners, irons, and electric kettles [32]. In the case of refrigerators, at least, they are not being enforced, so it remains to be seen if the new programme will fair any better than the existing scheme.

The four east European states with the closest ties to the EU (Poland, Czech Republic, Slovak Republic, and Hungary) are taking steps toward the adoption of the EU energy-labelling scheme. In part, this is motivated by a desire to facilitate the trade of appliances with the EU. Hungary was reported to be considering the independent introduction of MEES for refrigerators [33].

The UNECE have been conducting a study on appliance energy labelling and efficiency standards in Eastern Europe, known as the SACHA project. This is now being conducted by the Italian manufacturers association, ANIE, for the UNECE with partial funding from the SAVE programme of the European Commission. The original terms of reference of the UNECE project explicitly included the evaluation of mandatory minimum energy efficiency standards [34], however, these now appear to have been dropped without explanation.

6.2.1. Cold Appliance Standards in the Russian Federation

Domestic-refrigerator mandatory MEES were introduced as early as 1976 in the former Soviet Union and and were up-graded in 1984, 1988, and 1991 [35]. It has been claimed that these MEES have had the effect of approximately doubling the average energy efficiency of Russian- and CIS-made refrigerators since 1976 [32].

The 1976 minimum energy efficiency standards set upper limits to refrigerator energy consumption assuming that the refrigerator had a 1-star frozen-food compartment with a minimum useful volume of <7% of the total storage volume for refrigerators up to 180 litres; <9% for 181–300 litre refrigerators; and <10% for refrigerators with total volumes in excess of 300 litres. If the refrigerator had a 2-star frozen-food compartment its maximum allowable energy consumption had not to be more than 25% of that permissible for a 1-star model. Similarly, if the refrigerator had a 3-star or 4-star frozen-food compartment its maximum allowable energy consumption had not to be more than 50% of that permissible for a 1-star model. A separate class was also made for refrigerators with 'metal inner chambers', which were allowed to consume twice the energy of 'polymeric' models. If the volume of the frozen-food compartment exceeded that specified above, then an adjustment was made such that the permissible energy consumption was multiplied by a factor $(1 + E)$, where E is a coefficient dependent on the freezer volume defined as:

$$E = \frac{1.25}{100}\left(\frac{V_{FFC}}{V_{Total}}100 - A\right)$$

where V_{FFC} is the volume of the frozen-food compartment, V_{total} is the total storage volume, and A is the minimum volume of the frozen-food compartment (i.e. 7%, 9% or 10%, as described) [35].

The 1984 amendment to these standards made a distinction according to the quality of the refrigerator – Soviet refrigerators were classified and priced according to their quality by government planning bodies – such that their strictness was increased by 25% for the highest-quality models and 10% for 'first-quality' refrigerators (Table 7). The proportion of models that were in each class is not clear.

The energy consumption of Russian refrigerators is measured according to the standard GOST 16317, which also includes the minimum energy efficiency specifications. Up until 1987, this standard appears to have conformed to the ISO energy test standard for tropical-class models (i.e. using an ambient test temperature of 32°C); however, from this date the standard was adjusted to record energy consumption at the lower ambient temperature of 25°C, conforming to the ISO test for temperate-class models that is commonly applied in Western Europe. It is not clear how the MEES level was adjusted to reflect this change [35].

The threshold for the minimum energy efficiency standards was revised again in 1991 (see Table 7), but most of the Russian and CIS models failed to meet it. The majority of domestically produced models in the years 1991 to 1993 successfully met the 1984 standard for first-quality models.

An average Western European, 1992-make, 1-star refrigerator with 133 litres total volume and a 9.3 litre frost-box would consume 280 kWh/year under the EU energy test protocol EN153 [28]. The Russian 1991 minimum energy efficiency standards are set at a level roughly equivalent to this for one-door models; therefore, a significant proportion of the 1992 Western European models would fail to meet them. However, a typical Western

European, 1992-make, 4-star refrigerator-freezer with a 171 litre refrigerator compartment and 86 litre freezer compartment would consume 591 kWh/year under EN153 [28] and would better the 1991 Russian minimum energy standard of 780 kWh/year by 24%. Direct comparisons with the 1984 Russian standards are not possible because of the higher ambient test temperature in use compared to EN153; however, it is probable that almost all the 1992 Western European models would meet them. [35]

TABLE 7. Minimum Energy Efficiency Standards for Russian Federation 1-star Vapour Compression Type Refrigerators [35][1]

Total Volume (litres)	**1976 Standard**[2]	**1984 Standard**[2]		**1991 Standard**[3]	
	All Models (kWh/yr)	First-Quality (kWh/yr)	Highest-Quality (kWh/yr)	Single-Door (kWh/yr)	Double-Door (kWh/yr)
40	446	–	–	–	–
60	–	398	329	–	–
80	493	420	351	–	–
100	519	442	369	–	–
120	548	460	384	–	–
140	577	493	409	278	–
160	603	515	431	–	–
180	628	533	431	303	376
200	657	566	471	–	–
220	–	599	497	321	402
240	731	625	519	333	394
260	–	657	548	332	398
280	–	672	573		409
300	840	701	599	–	416
350	–	774	661	–	438
400	1023	847	723	–	566

[1] Separate standards exist for absorption-type refrigerators. The associated standards for 2- and 3-star/ 4-star refrigerators are described in the text.

[2] The energy value given by the standard is to be compared against the measured rating under GOST 16317-1976 (i.e. with an ambient test temperature of 32°C).

[3] The energy value given by the standard is to be compared against the measured rating under GOST 16317-1987 (i.e. with an ambient test temperature of 25°C).

6.2.2. Cold Appliance Standards in Hungary

Hungary introduced the EU cold appliance energy efficiency label in early 1995 [35]. At that time it was reported that the government was considering a proposal to introduce mandatory minimum energy efficiency standards designed to eliminate models in the least-efficient 'G' category of the label (see Section 7). Hungarian manufacturers were reported to be adapting their products to satisfy this goal [33]; however, this development is likely to be superseded by the MEES Directive in the EU as Hungary is beginning to harmonise their legislation with EU legislation as a prelude to membership. The same is likely to be true for Poland and the Czech Republic.

7. The European Union Energy Label

Unlike for minimum efficiency standards a framework energy labelling Directive was enacted in 1992 (92/75/EEC, date 22.9.1992) which granted the Commission authority to issue energy labelling implementation directives for cold-appliances, washing machines, clothes-dryers, dishwashers, ovens, water-heaters, lighting sources, and air-conditioning appliances. Other household appliances can be added to this list if significant energy savings can be identified.

Since that time implementation labelling directives have been issued for cold-appliances (92/2/EC, date 21.1.94), clothes-washers (95/12/EC, date 23.5.95), clothes-dryers (95/13/EC, date 23.5.95), washer-dryers (96/60/EC, date 18.10.96) and dishwashers (97/17/EC, date 16.4.97) while a energy labelling directive for household lamps was approved on 27.1.98 and will be published in the Official Journal of the European Communities in the coming days. Studies are underway to assess the technical basis of defining energy labels for electric storage water-heaters and air-conditioners, also a study has been completed which defines a methodology for the energy labelling of commercial refrigerated- and frozen-foodstuffs storage cabinets.

7.1 THE FORM OF THE ENERGY LABELS

The EU appliance energy-labels all follow a similar form. The energy efficiency of a given appliance is ranked into one of 7 bins graded from A to G. The information is presented with the seven efficiency classes (A to G) shown as vertically stacked horizontal-pointed, coloured bars. The stack of bars is always aligned to the left margin of the label with the A bar being at the top and the G bar at the bottom. The length of the bars increases incrementally from A to G while above the A bar is written *More efficient* and below the G bar *Less efficient*, so the respective length of the bars is intended to enforce the message that A means lower consumption (shorter) while G means higher consumption (longer). Similarly, the G bar is coloured bright red while the A bar is coloured deep green, while the other bars are coloured in a progression between the two hues. The colour scheme is again chosen to enforce the efficiency message as red is associated with danger and green with being environmentally friendly. The efficiency of the specific appliance model is indicated by attaching a larger black arrow that is aligned to the right and points back at the stacked efficiency class pointed-bars. The efficiency class bar it points to indicates the efficiency of the model.

In addition, to indicating the appliance efficiency the energy labels for the larger appliances contain a number of other items to indicate product performance. For cold appliances the energy consumption under standard test conditions is shown, expressed as kWh/year, the net storage volume of the refrigerator and the freezer compartments are indicated and the star-rating of the freezer compartment is shown. The compressor noise in dB is an optional extra although some EU countries require this by national law. See Figure 7.

For clothes-washers the energy consumption is given as the kWh/wash-cycle tested under a standard 60°C wash. Although most European consumers use lower wash temperatures (40°C is most common), appliance manufacturers prefer to test at 60°C so that the tests are more reproducible while it is argued that the results are directly proportional to the energy consumed under 40°C wash-cycles in the same machines. One of the most interesting features of the EU clothes-washer energy-label is that the cleaning and spin-drying performance of the machines are also indicated using two separate A to G scales. The format used is a horizontal listing of the performance scale written A B C D E F G where the model performance ranking is highlighted in bold text, e.g. A B C D E **F** G. The EU energy label is the only label which has such an explicit link between the appliance energy efficiency and it's performance. The spin speed, water consumption and washing capacity are also compulsorily indicated on the label, while noise may be indicated at the manufacturer's discretion; see Figure 8.

For clothes-dryers the energy consumption is given in kWh/drying-cycle measured under Norm EN 61121. The capacity and noise (discretionary) are shown and whether the dryer is an air-vented or condenser type is indicated.

The washer-dryer label shows both the energy consumption in kWh/wash-cycle tested under a standard 60°C wash (for the wash-only mode) and also shows the energy consumption in kWh/wash-dry cycle to wash and dry a full capacity wash-load at 60°C. The washing performance is indicated in the same manner as the clothes-washer label, ranked from A to G. Spin-speed (rpm), washing capacity (kg), drying capacity (kg of cotton wash) and water consumption (litres per full wash-cycle) are shown. The noise is again discretionary for the manufacturers and has a separate entry for the washing, spinning and drying cycles.

The dishwasher label is reminiscent of the clothes-washer label. Energy efficiency if grade from A to G on the colour bar scale but wash performance, and drying performance are also graded from A to G using emboldened horizontal lettering. Energy consumption is indicated in units of kWh/wash-cycle using a large font while the number of standard place settings, water consumption in litres/wash-cycle and noise (dB) are shown in smaller fonts.

The appliance brand and model number are clearly indicated on all the above energy-labels.

Energy labels for household lamps have posed a unique problem because lamps are generally small, low value items compared to white goods. Additionally, there is a huge amount of trade and resale of lamps across European boundaries, which has meant that it was not practical to use a label comprised of two parts that are need to be separately attached to the product. In response to this the Commission proposed a greatly simplified label that resembles the energy labels for white goods but has much less text and information in order to enable a language non-specific label to be produced.

The Commission has dispensed with applying a separately produced label to the product but will require the label to be printed on the lamp packaging. The label must be printed in colour if the lamp packaging is polychromatic but can be monochrome if the packaging is also monochromatic. The word *energy* is written at the top of the label with a variety of endings that spell energy in 9 of the 10 EU languages. The Greek spelling is written above the main word. In addition to indicating the lamp's energy efficiency from A to G the label also indicates the light output in Lumens, the rated power in watts, and the average lamp lifetime in hours. The problems of language means that this label even more than the other EU labels is a compromise between clarity and practicality. It is hoped that at least some consumers will know that the Lumen, Watt and h terms refer to light output, rated power and average lifetime respectively although this has not yet been tested. It is also hoped that the high purchasing rate of lamps will greatly add to consumer recognition of the energy label and that this will help to reinforce the message from all the labels.

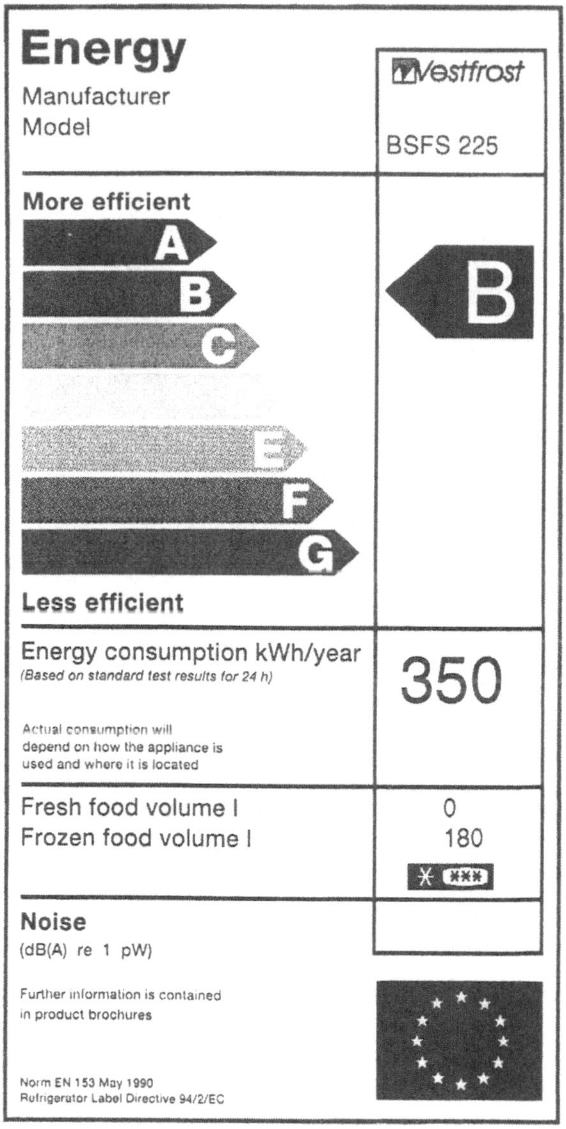

Figure 7.
The EU Cold Appliance Energy Label

Figure 8.
The EU Clothes Washer Energy Label

7.2 CALCULATING EU ENERGY-LABEL EFFICIENCY CLASSES

For cold-appliances the efficiency class grades A to G are defined for each product category in terms of an energy efficiency index I (defined as the tested model electricity consumption divided by the average model consumption for a unit of the same adjusted volume) as given in Table 8.

Table 8. Relative efficiency grades used in the EU energy label for cold-appliances

Energy Efficiency Index I	Energy Efficiency Class
$I < 0.55$	A
$0.55 \leq I < 0.75$	B
$0.75 \leq I < 0.90$	C
$0.90 \leq I < 1.00$	D
$1.00 \leq I < 1.10$	E
$1.10 \leq I < 1.25$	F
$1.25 \leq I$	G

Thus, a G class model uses 125% or more energy of an average cold-appliance of the same type and the same adjusted volume, while an A class model uses less than 55% of an average appliance of the same type and adjusted volume. The energy consumed by an average appliance of a given type and adjusted volume is calculated from the appropriate average performance reference line equation given in Table 3 and based on the values measured under the norm EN 153.

The efficiency spread indicated in these classes is based on several considerations. Firstly, that the efficiency of models on the EU cold-appliance market tends to be distributed[15] like a slightly skewed gaussian function except that there is a long tail on the side of the inefficient models while the efficiency of the most efficient model is bounded. Secondly, that the aim of the label is to encourage people to buy more efficient appliances - this means that it is not necessary to set the width of the efficiency classes such that there are an equal number of models in each class, but rather in such a way that it is a compromise between reflecting the real efficiency distribution of the market at the time that the label is introduced and the efficiency distribution one may hope to achieve after some years. Thirdly, that the highest efficiency class needs to be theoretically attainable even if it is not attained by any model at the time the label is introduced.

[15] The efficiency distribution refers to the graph one would see if one plotted a frequency distribution showing the number of models (on the y-axis) within any bracket of the energy efficiency index, I, on the x-axis.

When the cold-appliance energy label was introduced in 1994 there were almost no models in the A efficiency class available on the market; however, a detailed engineering analysis had shown that it was quite possible to manufacture A class models. Since that time several hundred A class models have become available on the EU market, justifying the principle of leaving the top efficiency class almost empty.

For clothes-washers the efficiency class grades A to G in terms of their specific energy consumption (kWh/kg) for a fully-laden 60 °C wash cycle as given in Table 9 and as measured under norm EN 60456.

TABLE 9. Relative efficiency grades used in the EU energy label for clothes-washers

Energy consumption 'C' in kWh per kg washed for standard 60 °C cotton cycle using the CENELEC harmonised test standard	Energy Efficiency Class
$C \leq 0.19$	A
$0.19 < C \leq 0.23$	B
$0.23 < C \leq 0.27$	C
$0.27 < C \leq 0.31$	D
$0.31 < C \leq 0.35$	E
$0.35 < C \leq 0.39$	F
$0.39 < C$	G

For the clothes-dryers the efficiency class grades A to G in terms of their specific energy consumption (kWh/kg) for a dry cotton drying cycle as given in Tables 10a and 10b and as measured under norm EN 61121. A distinction is made between air-vented dryers and condensing dryers because the air-vented dryers provide a slightly lower service as the moisture extracted from the clothes remains in the vented air for these dryers while it is condensed by a dehumidifier for the condenser dryers. On average the air-vented dryers have slightly lower energy consumption and are typically 36% less expensive than the condenser dryers.

TABLE 10A. Relative efficiency grades used in the EU energy label for air vented clothes dryers

Energy consumption 'C' in kWh per kg of load for a 'dry cotton' drying cycle, determined using the CENELEC harmonised test standard	Energy Efficiency Class
$C \leq 0.51$	A
$0.51 < C \leq 0.59$	B
$0.59 < C \leq 0.67$	C
$0.67 < C \leq 0.75$	D
$0.75 < C \leq 0.83$	E
$0.83 < C \leq 0.91$	F
$0.91 < C$	G

TABLE 10B. Relative efficiency grades used in the EU energy label for condensing clothes dryers

Energy consumption 'C' in kWh per kg of load for a 'dry cotton' drying cycle, determined using the CENELEC harmonised test standard	Energy Efficiency Class
$C \leq 0.55$	A
$0.55 < C \leq 0.64$	B
$0.64 < C \leq 0.73$	C
$0.73 < C \leq 0.82$	D
$0.82 < C \leq 0.91$	E
$0.91 < C \leq 1.00$	F
$1.00 < C$	G

For washer-dryers the efficiency class grades A to G in terms of their specific energy consumption (kWh/kg) for a complete fully-laden 60 °C wash cycle and dry cotton drying cycle as given in Table 11 and measured according to norm EN 50229.

TABLE 11. Relative efficiency grades used in the EU energy label for washer-dryers

Energy consumption 'C' in kWh per kg complete operating (washing, spinning and drying) cycle using standard 60 °C cotton cycle, and 'dry cotton' drying cycle, determined using the CENELEC harmonised test standard	Energy Efficiency Class
$C \leq 0.68$	A
$0.68 < C \leq 0.81$	B
$0.81 < C \leq 0.93$	C
$0.93 < C \leq 1.05$	D
$1.05 < C \leq 1.17$	E
$1.17 < C \leq 1.29$	F
$1.29 < C$	G

For dishwashers the efficiency class grades A to G in terms of the energy consumption per standard wash cycle (kWh/cycle) based on norm EN 50242. Both the dish cleaning and dish drying performance are ranked from A to G. The number of standard place settings and water consumption are also indicated on the label. The energy efficiency is determined by defining a reference energy consumption value C_R from:

$$C_R = 1.35 + 0.025 \times S \quad \text{when } S \geq 10$$
$$= 0.45 + 0.090 \times S \quad \text{when } S \leq 9$$

where S is the capacity of the appliance in standard place settings. An energy efficiency index E_I is then defined as:

$$E_I = \frac{C}{C_R}$$

where C is the energy consumption of the appliance. The efficiency classes are then given according to Table 12.

TABLE 12. Relative efficiency grades used in the EU energy label for dishwashers

Energy efficiency index, E_I	Energy Efficiency Class
$E_I < 0.64$	A
$0.64 \le E_I < 0.76$	B
$0.76 \le E_I < 0.88$	C
$0.88 \le E_I < 1.00$	D
$1.00 \le E_I < 1.12$	E
$1.12 \le E_I < 1.24$	F
$E_I \ge 1.24$	G

For household lamps the energy efficiency of a lamp is classed as A if:
for **Fluorescent lamps** without integral ballast (i.e. those requiring a ballast and/or other control gear to connect them to the mains)

$$W \le 0.15 \times \sqrt{\Phi} + 0.0097 \times \Phi$$

And for **other lamps** if:

$$W \le 0.24 \times \sqrt{\Phi} + 0.0103 \times \Phi$$

where Φ is the lumen output of the lamp and W is the power input into the lamp in watts.

If a lamp is not in class A then a reference wattage W_R is calculated from:

when $\Phi > 34$ lumens

$$W_R = 0.88\sqrt{\Phi} + 0.049 \times \Phi$$

when $\Phi \le 34$ lumens

$$W_R = 0.2 \times \Phi$$

The energy efficiency index of the lamp is then set as:

$$E_I = \frac{W}{W_R}$$

where W is the power input into the lamp in watts. The energy efficiency classes are then set as indicated in Table 13.

TABLE 13. Relative efficiency grades used in the EU energy label for household lamps

Energy efficiency index, E_I	Energy Efficiency Class
$E_I < 0.60$	B
$0.60 \leq E_I < 0.80$	C
$0.80 \leq E_I < 0.95$	D
$0.95 \leq E_I < 1.10$	E
$1.10 \leq E_I < 1.30$	F
$E_I \geq 1.30$	G

The household lamps directive does not cover low voltage halogen lamps or reflector lamps both of which have growing market shares in the EU.

All the product energy efficiency classifications used in the labels described have been chosen as a compromise between the desire to reflect the efficiency distribution in energy performance of models on the market at the time the labelling structure is introduced and the goal of encouraging the market to move into the higher efficiency classes. The selection of limits for the upper efficiency classes in the EU has generally been chosen in the light of a comprehensive product engineering analysis to determine the scope of untapped cost-effective high-efficiency design modifications. In the case of the cold appliances, clothes-washers and clothes-dryers for example, there were virtually no models in the A class when the labels were first introduced.

8. Voluntary agreements in the EU and Switzerland

8.1. WET APPLIANCE AGREEMENTS IN THE EU

In December 1993 the Commission contracted members of the GEA to conduct a detailed analysis of wet appliances. The study, which was completed in June 1995, found that [17]:

The potential for energy savings by proven technological improvements identified as being economically cost-effective shows that the efficiency of the average new European machine can be improved

- *by 25% for washing machines*
- *by 33% for dishwashers*
- *by 10% for dryers (clothes)*

Taking this study as it's starting basis, the Commission have set about negotiating a set of EU-wide voluntary energy efficiency targets with the European manufacturers' association CECED [26].

In the case of the clothes-washers, manufacturers have proposed to stop producing and importing domestic washing machines which belong to the energy efficiency classes E, F and G after 31.12.1997, although domestic washing machines with load capacity less than or equal to 3 kg which belong to class E are still to be permitted. As a second step, manufacturers have agreed to stop producing and importing domestic washing machines in the label classes D, E, F and G after 31.12.1999 for all machines except D class machines with a load capacity less than 3 kg or with a spin speed lower than 600 rpm. This last requirement is not based on any technical limitations of low spin-speed machines, but is a commercial compromise as most low spin-speed machines are produced for the southern European market and tend to be lower cost and lower efficiency. The link between the labelling Directive and the proposed minimum efficiency levels is explicit and is favoured by manufacturers. In practice, the first step in this agreement is already the de facto reality as there are very few 1997 made machines in these categories, see Figure 9. It is also of interest to note that there appears to be no connection between cleaning performance, which is also ranked on a scale from A to G on the energy label, and the energy performance. At least three manufacturers are now making triple A machines which have A energy performance, A cleaning performance and A spin-drying performance.

The Commission has demanded slight improvements on these thresholds and an agreement is thought to be close to being signed.

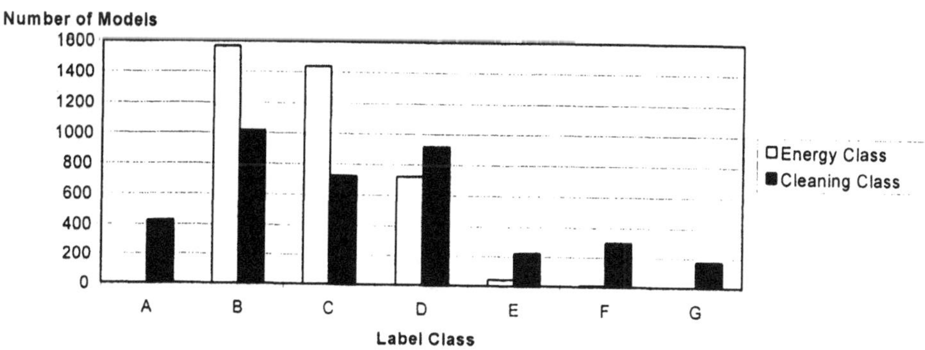

Figure 9.
Energy and cleaning performance distribution of EU manufactured clothes-washers in 1997: based on data from CECED

The Commission is hoping to negotiate similar voluntary agreements for clothes-dryers and dishwashers on the basis of the technical assessment of these appliances conducted by the GEA [17]. However, here there are greater technical problems. The energy label for dishwashers was delayed because of problems concerning the status of the energy test, which is still waiting approval, while the current range of clothes-dryers are not well spread between the efficiency classes of the energy label. This last point reflects the fact that it is difficult to achieve a high efficiency clothes dryer without using a much more expensive technology, such as a heat pump. Clothes-dryers are one of the few appliances discussed in this paper where there is a strong correlation between efficiency, technology and price and this is inhibiting efforts to establish a voluntary efficiency target.

8.2. STANDBY CONSUMPTION FOR BROWN GOODS

Following a SAVE sponsored study on the standby mode in brown goods [18] negotiations took place between EACEM (the European Association of Consumer Electronics Manufacturers) and the European Commission to agree voluntary limits for the stand-by mode consumption of TVs and VCRs sold in the EU. The aim was to agree to reduce consumption below the current average levels of 8.5W and 9.5W respectively. In 1997 an agreement was reached that stipulates that TVs and VCRs sold after the year 2000 will have a maximum standby consumption of 10W and that the sales-weighted average for each company signing the agreement will be lower than 6W. In addition, industry has committed themselves to pursue further activities to reduce standby power consumption in appliances.

8.3. SWISS TARGET VALUES

Outside the EU, Switzerland has adopted a 'target values' approach that combines elements of mandatory minimum energy efficiency standards and voluntary agreements with manufacturers [35]. In December 1990 the Swiss Federal Parliament adopted the 'Decree on the Use of Energy', which gives the Swiss Federal Office of Energy the power to issue requirements concerning the energy consumption of electrical appliances. Parliamentarians stated that mandatory energy efficiency standards should not be introduced unless the consumption values of appliances on the market failed to attain certain goals (target values) issued by the government for set dates in the future [36,37]. The first target value was applied to domestic refrigerators, freezers, and their combinations and commenced in the beginning of 1994; details are given separately below. Other appliances for which target values have been set are: electric ovens, dishwashers, washing machines, tumble dryers, TVs, VCRs, and office equipment (printers, computers, monitors, copiers, and fax machines). The target values for these appliances and their introduction dates are listed in Appendix B.

There are several important reasons why the Swiss approach is not the same as using either voluntary or mandatory standards. The target value system does not set a standard that all models must satisfy. Rather the target applies to the average of the whole new sales-weighted stock. This means that models less efficient than the target value can continue to be sold provided that the average of the stock satisfies the target. This approach necessitates a complex evaluation process because the Swiss authorities must track all the models sold on the market to understand if the target has been met. As a result, there is always a lag involved in gathering and processing the sales data and understanding the state of the market. Furthermore, because the manufacturers supply the data confidentially and at source, there could be verification problems. It is not made clear how multiple manufacturers divide responsibility for the efficiency transformation of the whole new appliance stock, especially given that they cannot know the efficiency-weighted sales of models in advance [35].

A crucial distinction between the Swiss approach and the type of voluntary agreement currently favoured by the EU Commission is that the target values are backed up by the existing mandate to impose mandatory efficiency standards at will. Manufacturers supplying the Swiss market are therefore fully aware that mandatory standards can be imposed without seeking further political approval if the target value approach fails. Attempts to negotiate voluntary agreements in EU states in the past have always lacked this vital element and have consequently always broken down.[16][35]

8.3.1. Cold appliance target values in Switzerland

Switzerland introduced a scheme for refrigerator energy efficiency target values in 1994 [35]. As in the EU energy efficiency standards proposal, the Swiss target values were determined following a statistical analysis of the refrigerator market. The product categorisation and adjusted volumes are very similar to those of the European Union. However, the Swiss Federal Office of Energy did not identify any significant statistical difference between the behaviour of 1-, 2-, and 3-star refrigerators and refrigerators without a low temperature compartment when their energy consumption's are expressed in terms of their adjusted volume. (Switzerland uses the same adjusted volume calculation and energy test protocol as the EU). The 1994 implementation deadline was actually a soft target, as the Swiss FOE say that 'it is intended from 1996 onwards that most of the appliances available on the market are to meet the consumption target values' [36].

In January 1995, about 49% of refrigerator-freezer models available on the Swiss market met or surpassed the Swiss target value, about 49% of refrigerators satisfied the Swiss target value, about 58% of upright freezers satisfied the Swiss target value, and about 49% of chest freezers satisfied the Swiss target value [36]. The Swiss target values are slightly tougher than the EU MEES requirements set out in Table 3.

9. The Savings Potentials of Appliance Energy Efficiency Standards in Europe

The potential electricity and CO_2 savings from the enactment of policies designed to shift the normal efficiency of the domestic appliance market toward the least life cycle cost (i.e. the minimum of the life cycle cost curve) are impressive. Furthermore, by defining the minimum energy efficiency standards to coincide with the point of least life cycle cost, it is intrinsically assumed that manufacturers preserve their profit margins while these savings are realised. In actuality, the profit per appliance sold

---

[16] The West German voluntary scheme of the late 1970s and early 1980s is the only successful example but this could not be renewed after a few years.

is anticipated to increase because it is assumed the purchase cost of the unit increases while the profit margin per unit cost remains constant.

In the opinion of the author, the goal of appliance energy efficiency policy ought to be to move the market toward the position of least life cycle cost. The least life cycle cost is itself a moving target as new technology and manufacturing cost relationships evolve. Thus, another area of policy can be to encourage the accelerated movement of the position of least life cycle cost toward higher energy efficiency. In practice, these levels of market transformation can only be brought about by the introduction of rolling mandatory minimum energy efficiency standards, preferably linked to market pull policies such as technology procurement programmes, etc.

9.1. COLD APPLIANCES

The GEA's analysis of the technological-economic options for cold appliances found that the least life cycle cost coincided with energy consumptions of between 45% and 62% of the average consumption of 1992 models depending on the cold appliance type [28]. Table 14, shows efficiency design options, estimates of their associated costs, savings implications and impact on the overall product life cycle costs.

TABLE 14. Life cycle cost and pay back periods of combined design options for 2-Star refrigerators [28]

Design options combined in order of increasing simple pay back period (The least life cycle cost design option is in bold)	Purchase Price (ECU)	Energy Consumption (kWh/yr)	Life cycle operating costs (ECU)	Life cycle cost (ECU)	Simple pay back period (Years)
0 = basecase model	470	335	386	856	-
1 = 0 + increased door insulation (+15 mm)	476	299	345	821	1.2
2 = 1 + decreased door leakage	477	293	338	814	1.2
3 = 2 + optimised compressor	487	258	297	785	1.7
4 = 3 + increased cabinet insulation (+15 mm)	509	204	235	744	2.3
5 = 4 + increased door insulation (+15 mm)	515	191	221	735	2.4
6 = 5 + increased cabinet insulation (+15 mm)	**536**	**166**	**192**	**728**	**3.0**
7 = 6 + doubled evaporator heat capacity	547	162	186	734	3.4
8 = 7 + doubled condenser heat capacity	556	159	183	739	3.7
9 = 8 + doubled condenser surface area	595	155	179	774	5.3

Figure 10 shows the forecast energy consumption of the entire EU stock of cold appliances under various policy actions up to the year 2020. These scenarios are: the static efficiency scenario (no improvement in cold appliance efficiency after 1995); the basecase scenario (this corresponds to the impact of energy labelling and natural market forces only); the EU MEES-Directive scenarios where 'the -15% standard' is introduced in either 1998, 1999 (as will happen) or 2000; the EU MEES-Directive scenarios followed by a second round of standards designed to improve average new model efficiency levels by 20% introduced in either 2002, 2003 or 2004; and a second round of efficiency standards introduced in 2002 set at the position of least life cycle cost.

Average, minimum and maximum cumulative electricity consumption and electricity savings estimates for these scenarios are given in Table 15.

TABLE 15. Forecast cold-appliance energy consumption within the EU under different policy scenarios [16]

Scenario	Cumulative consumption from 1996 to 2020 (TWh)			Cumulative energy savings from 1996 to 2020 (TWh)		
	Average	Minimum	Maximum	Average	Minimum	Maximum
Static efficiency from 1995	2708	–	–	(-278)	–	–
Base case	2430	–	–	0	–	–
-15% in 1998	2299	2344	2253	131	86	177
-15% in 1999	2311	2354	2267	119	76	163
-15% in 2000	2323	2364	2281	107	66	149
-15% in 1998/extra –20% in 2002	2169	2248	2090	261	182	340
-15% in 1999/extra –20% in 2003	2194	2270	2118	236	160	312
-15% in 2000/extra –20% in 2004	2218	2290	2146	212	140	284
15% in 1999/LLCC in 2002	1949	–	–	481	–	–

Large savings are forecast from the energy labelling of cold appliances compared to a static efficiency scenario[17] although the average market efficiency is unlikely to be static even without orchestrated government actions as normal market pressures lead to some efficiency improvements. The middle line marked '15% in 1999' shows the estimated impact of the MEES Directive (96/57/EC, date 18.9.96) that will come into effect at the national level on 3.9.1999. This is only thought likely to produce a relatively small saving compared to the labelling only scenario of 120 TWh by the year 2020 (some 4.9% of all projected cold appliance energy consumption over the same period). The bottom-most line shows the estimated impact of introducing mandatory minimum energy efficiency standards set at the position of estimated least-life cycle cost to come into effect in the year 2002. This would produce far larger cumulative savings of 481 TWh, worth 273 billion ECU in avoided bills, and some 1167 M-tons of CO_2 by the year 2020 (i.e. saving 19.8% of the labelling only scenario consumption over the same period) [16]. The intermediate scenarios show the impact of imposing a second round of minimum energy efficiency standards requiring an extra 20% efficiency improvement in the models offered for sale at three different dates after the first round of standards are introduced in 1999.

[17] Overall the combination of market forces and the energy label is forecast to result in electricity savings of 278 TWh over the period 1996 to 2020 compared with a frozen or static efficiency scenario. The value of this avoided-electricity to consumers is ~39 billion ECU (> US$40 billion) if one assumes constant real electricity prices [16].

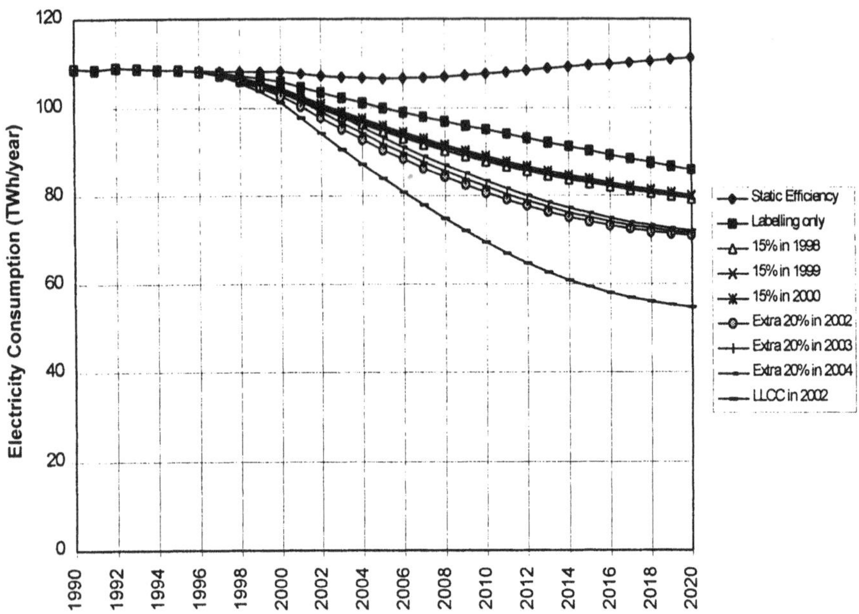

Figure 10.
Estimated and forecast cold-appliance electricity consumption in the EU for a range of labelling and standards policy scenarios [16].

9.2. WET APPLIANCES

For wet appliances, cumulative energy savings of 222 TWh by 2016 have been projected, assuming minimum efficiency standards or targets set at the ETP[18] (full Economic and Technical Potential) level were introduced [17].

If it is assumed that electricity prices remain fixed at 1994 values, the ETP scenario leads to saved electricity costs compared to the BAU of 28.8 billion ECU by 2016, Table 16. Additional running cost savings would be anticipated through reduced water bills, but these have not been included in the calculation. A NPV cost analysis has not yet been done for these appliances.

---

[18] The ETP is another way of saying the energy efficiency level coincident with the least life cycle cost.

TABLE 16. Cost Implications of Least Life Cycle Cost Mandatory Energy Efficiency Standards for Wet Appliances in the EU [17]

	Saved Electricity Consumption (TWh)	Saved Electricity cost to consumers (1994 billion ECU)	Avoided Carbon dioxide emissions (MT CO_2)
Annual Saving in 2016 (BAU-ETP)	19	2.4	9.8
Cumulative Saving (1995-2016) (BAU-ETP)	222	28.8	114

9.3. STAND-BY LOSSES

Reducing the standby mode consumption of domestic appliances could also result in large energy savings. Stand-by losses can typically be reduced by 75-80% compared to the current average values using cost effective design options that retain the stand-by functionality. Figure 11 shows a rough estimate of the annual average consumption in stand-by mode for domestic electric appliances in Western Europe (≈24 TWh/yr) assuming that the saturation of these appliances is the same as for Switzerland in 1990. More recent estimates suggest that the total energy consumption of domestic appliances in the standby mode could be more than double this value over Western Europe. Figure 11 also indicates what the consumption would be in the unlikely event of a uniform 100% saturation of all these appliances (≈90 TWh/yr). Implementing either voluntary or mandatory design changes could cut this usage by at least four-fold. For TVs and VCRs, it is estimated that standby loads of about 1W are technically feasible and economically justified which should be contrasted with typical consumptions of 9W and 15W for 1994 models of these appliances respectively [18]. Product design changes needed to achieve this saving are estimated to cost no more than ~20 ECU per appliance. This cost would be paid back in 2.2 years assuming average EU electricity prices and near continuous standby use.

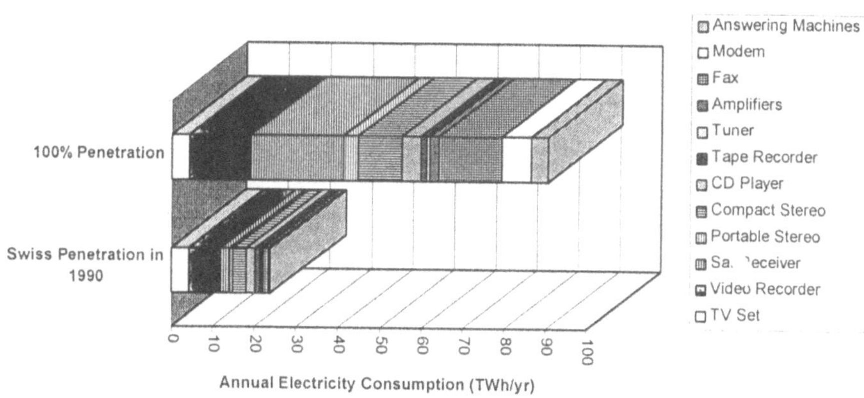

Figure 11.
Estimated Stand-By Consumption in the Residential Sector for Western Europe.[Author]

10. Conclusions

Given that underlying demand for energy services continues to increase in all consuming sectors it is imperative for the European Union and non-member states to aggressively pursue energy efficiency measures if they are to have any realistic chance of meeting their stated climate change commitments. The promotion of appliance energy-efficiency is a win-win policy in that total electricity consumption can be reduced with no loss of service or manufacturer profitability and at net financial benefit to the consumer. Global experience suggests that a judicious mix of policies, and especially mandatory energy labelling and minimum energy efficiency standards, can provide highly cost-effective and deep cuts in European electricity consumption. The adoption of a Framework Directive on appliance minimum energy efficiency standards could save European consumers billions of ECUs in the life cycle costs of their appliances while producing important CO_2 savings. For cold appliances alone, the setting of a mandatory minimum energy efficiency standard coincident with the point of least life cycle cost would save each person in the EU an average of ~61 ECU in net present value costs. Even though electricity and appliance costs vary appreciably from country to country within the EU, the point of least life cycle cost is comparatively invariant for many appliance types. As a result it is possible to set uniform standards across Europe for a range of appliances giving economic benefits in each region.

11. References

1. NF (1962) Économie Domestique Réfrigérateurs Ménagers A Compression, Norme Française D38-301, April.

2. NF (1978) Conservateurs Ménagers De Denrées Congeleés et Congélateurs Ménagers, Norme Française D38-320, September.

3. CENEf (1995) *Identifying Equipment for Energy Efficiency Standards: Draft Report*, Center for Energy Efficiency, Moscow, February.

4. UNECE (1995) Report on the meeting of the third session of the *Ad Hoc Group of Experts on Energy Efficiency Standards and Labelling Systems* of the Economic and Social Council of the United Nations Economic Commission for Europe, Held in Milan, 1-2 December 1994.

5. Bertoldi, P. (1995a) European commission efforts to promote more efficient use of electricity: the pace programme, Presentation made at the SAVE-Workshop 'Efficient Electric Appliances in the Domestic and Office Sector', October 12-13, Vienna.

6. de Almeida, A. T. (1995) Personal communication with expert conducting the study on industrial motors for the European Commission, Dept. Eng. Electritecnica, University of Coimbra, Portugal.

7. BRE (1994) *Study Of Measures To Promote Energy Efficient Lighting in the Commercial Sector In Europe*, Building Research Establishment, Garston, Watford, WD2 7JR, UK, for the Commission of the European Communities, Brussels, Belgium, Contract EC-DGXVII No. 4.1031/E/93-01, December.

8. EC (1996) *Energy in Europe: 1996 Annual Energy Review: Special Issue*, DG-XVII the European Commission, September.

9. IEA (1995) *Energy Policies of IEA Countries: 1994 Review,* The International Energy Agency, Paris.

10. CEC (1994) *Energy in Europe: Annual Energy Review,* DG-XVII (Energy), Commission of the European Communities, Brussels, April.

11. De Almeida and Fonseca (1997) 'Characterisation of the electricity use in European Union and the Savings Potential in 2010' A. de Almeida and P. Fonseca in *Energy Efficiency Improvements in Electric Motors and Drives* A. de Almeida, P. Bertoldi and W. Leonhard, Springer-Verlag, Heidelberg, 1997.

12. Hinnells, M.J. & Lane, K.B. (1995) The relative importance of technical and behavioural trends in electricity consumption by domestic appliances, *The Energy Efficiency Challenge for Europe: Proceedings of the 1995 Summer Study* Mandelieu, France, European Council for an Energy Efficient Economy, Stockholm, Sweden.

13. DGEMP (1993) *Tableaux des Consommations d'Énergie en France,* Edition 1993, Direction Generale de l'Energie et des Matières premières, Minstère de l'Industrie, des Postes et Télécommunications et du Commerce Extérieur, 75353 Paris, France.

14. DECADE (1996) *DECADE: Second Year Report 1995* (provisional results), Energy and Environment Programme of the Environmental Change Unit, University of Oxford, UK.

15. Schipper (1997) Lee Schipper, International Energy Agency, personal communication.

16. Waide (1996) *European domestic refrigeration appliances: minimum energy efficiency standards scenario analysis* PW Consulting for the French Energy and Environment Agency, ADEME, November.

17. GEA (1995) *Washing Machines, Dryers and Dishwashers: Background Reports Volumes I, II & III* analyses conducted for DG XVII of the European Commission, published by Danish Energy Agency, Copenhagen, June.

18. NOVEM (1995) *Study of standby losses and energy savings potential for Television and Video Recorder sets in Europe: Part A* NOVEM for DG-XVII of the Commission of thge European Communities, SAVE contract 4.1031/E/95-001.

19. Turiel, I., Koller and McMahon, J. *Overview of international energy standards* (Draft report), International Energy Agency, Paris, 1995.

20. EVA (1998) *Analysis of energy efficiency of domestic electric storage water heaters* Energieverwertungsagentur (EVA) for DG-XVII of the Commission of the European Communities, SAVE contract 4.1031/E/95-013, February.

21. Moeller J., & Nielsen B. (1991) *Ny Model for boligers elforbrug* (New Model for Household Electricity Use), Technical Report 281, DEFU, Lyngby, Denmark.

22. Sakulin M., & Dell, G. (1991) *Energiesparpotential elecktrischer Haushaltsgeräte,* Institut für Elektrische Anlagen, Technische Universtät Graz, Austria.

23. Appliance (1995) Portrait of the European Appliance Industry, *Appliance*, June.

24. Euromonitor (1994) *The World Market for White Goods: Strategy 2000* Volume 1, Euromonitor, London.

25. Euromonitor (1995) *European Marketing Data & Statistics 1995, 30th Edition,* Euromonitor, UK

26. Bertoldi, P. (1995b) Personal communication with the manager of the appliance energy efficiency programme of DG-XVII (Energy) of the Commission of the European Communities, Brussels, September.

27. Waide (1997) *Analysis of the energy efficiency of 1994 domestic refrigeration appliances: Part B: Long-term impact and scenarios analysis*, PW Consulting for TNO and DG-XVII of the European Commission, January 1997.

28. GEA (1993*) Study on Energy Efficiency Standards for Domestic Refrigeration Appliances* Group for Efficient Appliances (comprised of the following national energy and environment agencies: Ademe (France), Danish Energy Agency (Denmark), Novem (the Netherlands)) for DG-XVII of the European Commission, March.

29. OT3E (1994) *Energy-Efficient Office Technologies in Europe (OT3E)* Final Report, University of Bordeaux for the Commission of the European Communities, Contract EC-DGXVII N° 4.1031/E/92-07.

30. UoC (1996) *Actions to promotoe energy efficient electric motors*, Univeristy of Coimbra, Contract no. SAVE 4.1031/E/94-08 for DG-XVII of the Commission of the European Communities, October 1996.

31. Biktchentaev, A.A. (1994) The Russian Federation practice in the field of energy efficiency standards and labels', paper presented at the third session of the *Ad Hoc Group of Experts on Energy Efficiency Standards and Labelling Systems* of the Economic and Social Council of the United Nations Economic Commission for Europe, Held in Milan, 1-2 December 1994.

32. Bashmakov, I. (1995) Personal communication with the director of CENEf (Center for Energy Efficiency), Moscow.

33. Terstyansky, T. (1994) Personal communication with an official of the Hungarian Energy Office (Magyar Energia Hivatal), Budapest, Hungary.

34. UNECE (1993) *Comprehensive Report on Energy Efficiency Standards and Labelling Systems* Report N° ENERGY/AC.12/R.3 by the *Ad Hoc Group of Experts on Energy Efficiency Standards and Labels* of the Economic and Social Council of the United Nations Economic Commission for Europe, 10-12 May.

35. Waide, P. & Lebot, B. (1995) Refrigerators and Freezers: Market Characteritics, Energy Use and Standards Harmonisation: Draft Report prepared for the International Energy Agency and presented at the Experts Meeting on International Harmonisation of Energy Efficiency Standards, 30-31 May, Paris.

36. SFOE (1995) private communication and *Consumption Target Values for Household Appliances: Information Sheet*, *Consumption Target Values for Office Equipment, Television Sets and Video Recorders: Information Sheet*, and *Verbrauchszielwerte von Haushalt-, Büro- und Unter-haltungselektronikgeraten*: technical paper, Swiss Federal Office of Energy, Bern, Switzerland.

37. Bokämper, S. (1995) *Promotion of Energy Efficient Technologies in the European Union: The Case Of Refrigerators*, Masters Thesis, Technische Universität Clauthal Institut für Elektrische Energietechnik, Germany.

Appendix A. Appliance Energy Efficiency Programmes in Western Europe

Appliance Type	National Studies	EU/EC Studies	Energy Test	Energy Label	Eco Label	Rebates	Procurement	Standards
DOMESTIC								
Refrigerators	✓	✓✚	✓	✓	✓	v✚	v	✓▲♣ ☆
Freezers	✓	✓✚	✓	✓	✓	v✚	v	✓▲♣ ☆
Clothes Washers	✓	✓✚	✓	✓✠♦	✓		v	➲♠▲♣
Clothes Dryers	✓	✓	✓	✓✠		✚	v	▲
Washer-dryers		✓	✓	✓				➲
Dishwashers	✓	✓	✓	✓✠♦				➲♠▲
Electric Water Heaters	✓	✓	✓	✚♦		✚		✚♣
TVs	✓	✓→	□	♥✣♦				➲▲
VCRs	✓	✓	□	♥✣				➲▲
Microwave Ovens	✓	→	✓	✠				
Electric Hobs	✓	→	✓	→✠				
Electric Ovens	✓	→	✓	→✓ (old)				♣▲
Air Conditioners	✓	✓	✓	→✚				
Lamps	✓	✓	→	✓	→	v		
Ballasts			→			v	v	
Electric Heating	✓		✓					
Irons	✓		✓					
Kettles	✓							
Vacuum Cleaners	✓		✓					
COMMERCIAL								
PC	✓	✓	□❒	✣★	✿			➲▲
Monitors	✓	✓	□❒	✣★	✿		v	➲▲
Faxes	✓	✓	□	✣★			v	➲▲
Copiers		✓	□	✣★	✤		v	➲▲
Serial Printers	✓	✓	□	✣★	✿			➲▲
Laser Printers	✓	✓	□	✣★	✿			➲▲
Refrigeration	✓	✓	✓	✚			→	
Air Conditioning Systems	✓	→	✓	→✚			v	
Heat Pumps		✓	✓	→✚			v	
Ballasts		✓	✓				v	✚→
Lights	✓	✓	→	→			v	✚→
Ventilation Systems		✚					v	
Heating Systems								
Vending Machines	✓							
INDUSTRIAL								
Motors	✓	✓	✓					➲

Key: All symbols apply to the EU unless stated.

✓ = enacted

➡ = in progress
➲ = voluntary target value scheme in progress (➲ = under discussion)
✚ = proposed
v = country or region specific programme
♥ = GEA-VIA voluntary award scheme
▲ = Swiss voluntary target value scheme
□ = Swiss test method defined
❀ = Swiss ICMF eco label
❒ = Swedish test method defined
♠ = Danish mandatory energy efficiency standards proposed
✠ = German voluntary labels
♣ = German voluntary target value scheme (1978 to 1985)
✤ = German Blue Angel environmental labelling scheme
♦ = French mandatory labelling scheme since December 1976
✰ = French mandatory energy efficiency standards (1966/1978 to 1984)
❖ = Swiss E-2000 Energy award label
★ = Adaptation of USEPA Energy Star to include Europe & Japan (in progress)

Appendix B. Swiss Target Values [36]

Table B1. Consumption Target Values for Office and Entertainment Equipment [36]

Appliance	Operating Mode	Consumption Target Value	
		Max. Consumption (W)	to be met by
Work Station Computers[1]	On	Z=5	end of 1997
	On	Z=3	end of 1999
(with integrated monitor)	Sleep-mode	Z=15	end of 1997
(with integrated monitor)	Sleep-mode	Z=13	end of 1999
(without integrated monitor)	Sleep-mode	Z=10	end of 1997
Screens[2]	Sleep mode	Z=5	end of 1997
	Sleep-mode	Z=3	end of 1999
Printers[3]	Off	Z=1	end of 1996
	Stand-by	Z=2	end of 1996
Copiers[4]	Off	Z=1	end of 1996
	Stand-by	Z=27+3.23·c*	end of 1996
Telefax Machines[5]	Stand-by	Z=2	end of 1997
Television Sets[6]	Stand-by	Z=5	end of 1996
	Stand-by	Z=3	end of 1997
Video Recorders[7]	Stand-by	Z=6	end of 1996
	Stand-by	Z=3	end of 1998

Terms of Applicability:

[1] Computers which are used as desktop or tower models at work desks (PCs, work stations, microcomputers, etc.) including those with a monitor integrated in the same casing. Devices that can be operated independently of the mains and those supplied as servers are not included.

[2] Monitors sized up to 21" diagonally across.

[3] Printers with a printing speed of up to 25 pages per minute, including the mains adapter supplied by the manufacturer.

[4] Targets do not apply to machines with a speed over 100 copies per minute, or to black and white copiers for paper sizes larger than A3 or to full colour copiers.

[5] Telefax machines, including those with additional functions, such as integrated comfort telephone, answering machine, etc.

[6] Mains operated television sets with a visible screen diagonal of more than 20 cm, including auxiliary equipment such as remote control receivers, electronic clocks, etc. TV sets with integrated video recorder, and those equipped with an integrated satellite receiver, picture tube format of 16:9 or database interface, are not included.

[7] Non-professional video recorders including standard accessories supplied by the manufacturer. Digital video recorders are not included.

* c is the copying speed.

Table B2. Consumption Target Values for Four Household Appliances[36][#]			
Appliance	**Sub-Categories**	**Consumption Target Value**	
		Max. Consumption (kWh)	**to be met by**
Electric Ovens[1]	Ovens with exhaust fans	Z=0.9kWh per cycle	end of 1997
	Other electric ovens	Z=0.8kWh per cycle	end of 1997
Dishwashers[2]	Dishwashers for less than 8 place settings	Z=0.13kWh per setting	end of 1997
	Dishwashers for 8 or more place settings	Z=0.12kWh per setting	end of 1997
Washing Machines[3]	-	Z=0.23kWh per kilogram	end of 1997
Tumble Driers[4]	Air exhaust tumble dryers	Z=0.60kWh per kilogram	end of 1997
	Water condensation dryers	Z=0.65kWh per kilogram	end of 1997
	Air condensation dryers	Z=0.70kWh per kilogram	end of 1997

Terms of Applicability:

[1] Electric ovens for household use. Ovens with an integrated microwave cooker are not included.

[2] Electric dishwashers for household use.

[3] Electric washing machines for household use including those with an integrated tumble dryer.

[4] Electric tumble dryers for household use.

EFFICIENT TECHNOLOGY AND APPROPRIATE LIFE-STYLES

JØRGEN S. NØRGÅRD
Department of Buildings and Energy
Technical University of Denmark
DK-2800 Lyngby, Denmark

Introduction

The introductory contribution in this volume presents a broad view on the greenhouse effect and other reasons *why* we in Europe have to change dramatically our use of energy[1]. Especially the consumption of electricity should be kept in mind, since it constitutes the most polluting form of secondary energy, when produced from the presently most widespread primary energy sources.

Also it was suggested *how* electricity savings could be achieved by using more energy efficient end-use *technologies*. The volume has presented many examples of how this improvement in end-use efficiency can actually be implemented, and how much it can contribute to solving the environmental problems associated with the supply of electricity.

Here at the end of the volume, I will turn to other aspects of our societies than just technology. This is essential because we are not going to save the environment by changing path in development of technology only. It is decisive to redirect also the course of the way we organize our societies and daily lives in Europe and other Western industrialized societies. We must redefine the targets for the development towards a better and more direct satisfaction of all people's need within an environmentally sustainable economy.

1. Extending the Energy Chain

Over the past decade or two we have gradually come to recognize that *electricity* consumption should not be considered an end in itself, but rather a means to obtain *energy services*, as discussed in the earlier paper[1], illustrated there by the energy chain. We should, however, expand our view beyond technology by similarly recognizing, that energy services must not become an end in itself, but should also be considered a means only. Somehow we should be able to agree that the ultimate end, the objective of the development, should be a high quality of life for all human beings in a

S. Nørgård et al. (eds.), Development with Sustainable Use of Electricity, 287-300.

just and environmentally sustainable world. Obviously, energy services can contribute to a better life, but only to a certain extent. Being at a level of living standard like the average in Europe, people gradually change towards other preferences than just more material welfare, and this should be recognized and welcomed in an environmentally oriented policy, not least in the electricity sector.

Discussing these issues of welfare, standard of living, and lifestyles point towards adding an extra link to the energy chain from the first paper, as shown in Figure 1. In other words, how appropriate or "efficient" is the *lifestyle* we practice, or aim for, in turning the energy services and other materialistic inputs into human satisfaction? We will here interpret the life-style as the whole set or pattern of human behavior. As far as electricity consumption is concerned, lifestyle could be defined as the non-technical parameters which determine a person's consumption of electricity. As suggested by the shape of the lifestyle parameter in Figure 1, this is a "soft" issue in the sense that few hard data are available, and it often makes little sense to ask for them. This does, of course, not imply that such issues are of less importance than the technology, economy and other aspects which just happen to be quantifiable. In this chapter the lifestyle concept is in this paper interpreted in the broad sense, including for instance the economic policy setting the frame for the way people live.

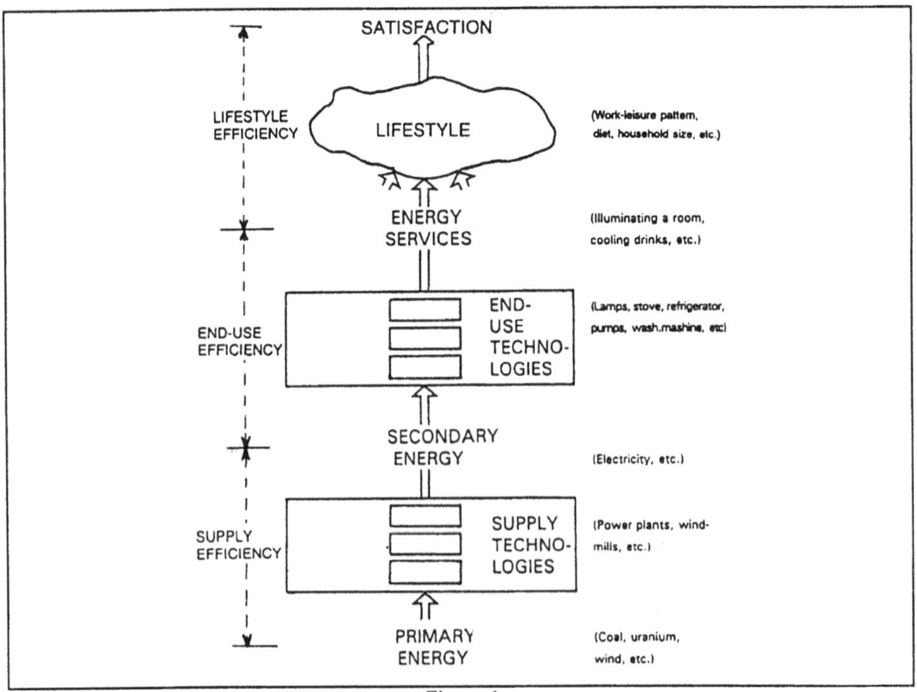

Figure 1.

The energy chain shown in the introductory paper[1] is here extended to illustrate that a high degree of satisfaction should be the ultimate goal of our development, not just as much energy service as possible.

1.1. FAULTS IN ENERGY POLICIES

Although environmental considerations, including energy aspects, have become daily issues in politics, they still are *not integrated* in decisions in all political sectors like tax policy, labour policy and economic policy in general. Furthermore, energy policy is still usually associated with the energy *supply* side only, and when finally *demand* side policies are implemented, they are often sending some wrong signals, as discussed in the following.

2. Efficiency Pitfall

One of the mistakes in present energy saving policies is that almost all efforts are directed towards improving *technological end-use efficiencies*. This implies a risk of falling into what can be termed the efficiency pitfall. As an example, the end-use efficiency of a refrigerator will often be measured by the electricity consumption per volume, such as annual kWh per litre of cooled storage volume. This efficiency can be improved by just choosing a larger refrigerator, leaving everything else like insulation of the cabinet and refrigeration system efficiency the same. The larger refrigerator will indeed consume more electricity. But because of some "economy of scale" effects like smaller surface to volume ratio, the larger unit will use less electricity per litre volume, and hence be termed more efficient. Similar effects occur in many other uses of energy. Heating or cooling a bigger house will usually be more efficient measured by energy per m^2 floor space, but nevertheless, totally it will consume more energy. Driving longer distances by car will of course take more gasoline. But because of better performance of the motor etc., it usually consumes less per km, and hence can be termed more efficient. In other words, in many cases we can improve the energy efficiency by consuming more energy!

If we really needed a larger refrigerator, a bigger house like 10 m^2 per person instead of 5, etc., then the overall efficiency would most likely have benefitted from the larger choices. But this is usually not the case when talking about the lifestyles presently conducted in the industrialized countries with 30 - 50 m^2 domestic floor space available per person. Because of general rules about diminishing returns from growing consumption, we will normally not get twice as high a degree of satisfaction when doubling the size of the refrigerator, the house, the daily travel to work, or other energy services. If the gains in technical end-use efficiency are achieved as described above, by increasing the energy services, they are obtained at the expense of lifestyle efficiency, see Figure 1. The result can be that the *overall efficiency*, including both the end-use technology and the lifestyle, have actually declined by this misled effort to improve energy efficiency.

The proper way to utilize the above described "economy of scale" effects is to practice *more sharing*. Larger households are one obvious path towards increasing lifestyle efficiency. But also for instance two families sharing a 300 litres chest freezer instead of each having a 150 litre model, will typically save 20-40% on the electricity used for that purpose. Other examples are the car pooling or the use of mass transport instead of cars.

Actually the efficiency pitfall here described for single end-uses actually also applies to an economy as a whole as discussed later. This calls for careful use of the technical efficiency as the sole target.

3. The Myth of Decoupling Economy and Energy

Before the oil shock in 1973, electricity demand in many parts of the world had been growing at a rate higher than that of the economy, typically 7% versus 4%. Afterwards, gradually the growth rates of electricity consumption, and of energy consumption as a whole, have been significantly lower than earlier and slower than the economic growth. This has been termed *decoupling* between the energy consumption and the economy, suggesting that no links exist between the two developments. It is highly questionable whether the phrase decoupling is reflecting the actual development.

Electricity consumption is, and will always be, in one way or another related to the production or operation of consumer goods. Consequently, the higher this economic activity is, the higher is the electricity consumption, *everything else being equal*. For instance, no matter how efficient a refrigerator is, two of them still consume twice the amount of electricity as does one. In other cases where the electricity consumption is more related to people's behavior and needs, as in the case of a TV set, two of them in a household will also consume more electricity than one TV set, although not necessarily twice as much.

During periods with technological or structural changes, as for instance the improvements of end-use efficiency, the coupling factor might undergo changes, and in macroeconomics this will appear as a decoupling. But if - or when - these changes in technology and social structure come to an end, the growth rates for the electricity consumption once again will catch up with that of the economy. The technical gains can buy us time, but they cannot resolve the basic conflict between environmental sustainability and eternal growth in the economy. The coupling between economic activity and electricity consumption is always there. Hence it is essential in a session on electricity savings also to take a closer look at the economic development.

4. Structural Efficiency Gains

As an economy develops, its structure changes in accordance with the changes in people's preferences as consumers. The economies in the highly industrialized countries are moving from a dominantly materialistic consumption towards more spiritual and human-related aspects of life. This shows up as a growth in the service sector, public as well as private service. To provide an ECU worth of child caring, hair cut or other service requires less energy than providing an ECU worth of steel, concrete or other manufactured goods. Furthermore, the industrial productions are also undergoing changes towards output that is more information intensive and less intensive in energy and other physical resources. All this is causing the overall energy efficiencies of the economies, measured as units of GDP per unit of energy, to increase. In other words, for a country's whole economy we are experiencing an overall *efficiency pitfall* as described above for single end-use technologies. By increasing a country's GDP, and usually also its *energy consumption*, the economy can be presented as becoming more energy efficient. This trend in development towards less energy-intensive economic activities has also led some politicians to suggest that economic expansion can go on forever, because the growth will occur within the service sector. There are more reasons why this can hardly become the case, as outlined below.

5. The Less Energy-Intensive Service Sector

If we apply an input-output table to the analyses of energy intensities of the economic sectors, the difference between the manufacturing sector and the service sector is not as marked as often assumed. The point is that all sectors are interconnected, and when one of the economic sectors expands it will in principle affect them all. Such input-output analyses of the energy- intensities of Danish economic sectors have been carried out by the economist Jespersen[7]. Surprisingly the result for Denmark was that one million ECU worth of GDP in the *private service* (or market service) sector, including trade, hotels, and transport, accounts for almost as much energy input as a similar production in the manufacturing sector, namely 6.9 TJ versus 8.4 TJ. Only what are termed non-market services, which have typically been *public service*, such as education, health, administration, child care, and elderly care, show a significantly lower energy intensity of 3.1 TJ per million ECU. This demonstrates that growth in the service sectors is not harmless as far as energy and environment is concerned. Especially the market service is only around 20% less energy demanding than the manufacturing sector. If looking at the demand for *electricity* only, the differences will probably be even smaller. The more human-related non-market service sector is clearly less harmful to the environment. Some gains can be achieved by redirecting the economic growth to this traditional public service sector. Such a switch, however, will not be brought about by market forces. It will require determined political actions, and presently most governments in Western Europe are actually trying to hold back on expansion of this sector. Growth in this sector is in the long run facing other problems, as discussed in the following.

6. Short-sighted Faith in Service Society

Politicians are often relying on the service sector to rescue their philosophy of continuos economic expansion from causing environmental problems. This, however, turns out to be somewhat short-sighted, because of the limited options for expansion in this sector, especially in the least harmful non-market service. Despite an obvious need for more of such person-to-person related services, it is a question to what extent such human relations will be professionalized and thereby contribute to GDP. Let us imagine that in the future a 3-4% economic growth per year in a country like Denmark could occur in this human-related economic sector only, with no input-output relations to other sectors. In this case, this service sector would have to expand to *12-15 times its present size* over the next 40 years. There seems to be no perspective in relying on economic growth in such professional caring, the overall efficiency of which might already have turned negative in some countries.

6.1. TRENDS IN VALUES AND LIFE-STYLES

Already back in the 1950s USA was facing problems about keeping up the economic growth due to a general saturation. Contrary to the European nations, USA came out of the Second World War with a rather intact production capacity, and no need for large rebuilding programs. Soon the inventories were full due to a general saturation among the consumers, who felt they were already rather well off. This was posing a threat to the overall economic growth strategy, and countermeasures were taken[3]. One was the introduction of psychology in advertising, appealing to the subconsciousness of the consumers. This "brainwashing" which has now become common practice, was permitted, despite of critique from religious groups and other organizations. They were protesting, not from an environmental point of view, but for the sake of human dignity, etc. Together with other actions, the psychological advertisement managed to get the economy growing again. Today, 50 years later, we have a new reason to welcome such saturations, namely the environmental cause.

7. Surveys Indicating Saturation

Several surveys indicate a growing public concern for the economic growth development, usually expressed through people's own preference in life. Figure 2 shows the development in the attitudes of people in Denmark, as they were expressed through a series of systematic surveys, carried out by the Danish National Institute of Social Research[4]. As seen, by the late 1980s, 70% of the Danes would prefer having more time with family, friends and private activities in general, rather than more income and consumption. This preference for issues not associated with the monetary economy, is what we here term a *general economic saturation*.

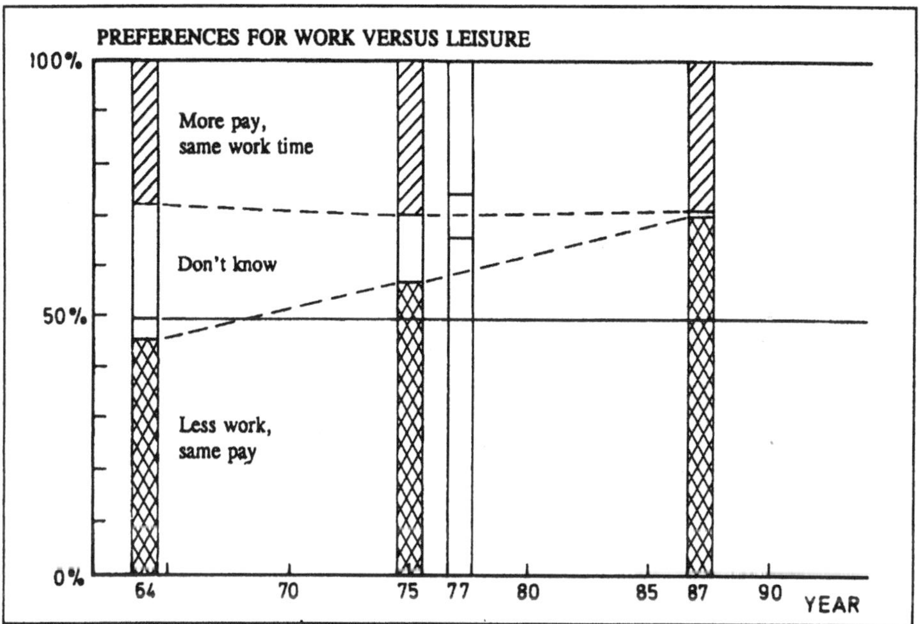

Figure 2.
Developments in attitudes towards leisure, work, and income in Denmark, according to surveys of people's preferences.

Other surveys in Denmark and North European countries in general show similar trends. After 1987 other surveys with slightly different questions asked than in the ones referred to above, have shown more blurred trends in Denmark with preference for more leisure, going down and up.

It is worth noticing, that these trends towards economic saturation were reached in a period in which consumers were increasingly subjected to advertisements on TVs, streets, mail, etc.

8. Working Patterns

The result of the above described surveys is somewhat confirmed by real actions. As a means for reducing unemployment, thousands of workers in Denmark and Germany have been offered - and have accepted - reduced working time. Their decline in working hours has only partly been economically compensated by some increase in hourly pay, so in general their incomes have been lowered. This demonstrates that one of the most deep-seated values in the North Europeans countries, the appreciation of hard work as a value in itself, is giving way.

This could pave the way for new trends in economic policy. Presently the main argument for increasing production is *not* the need for the products, but the need to "create jobs". Following the trends in people's attitudes, this argument could be abolished in the future.

Job sharing and reduction in working hours is probably the most important contribution a government can make to a sustainable development which satisfies people's needs and wants with a low consumption of electricity and resources in general.

8.1. TRENDS IN ECONOMIC POLICY

In this section we will look at the trends in the dominating economic policies and discuss its consequences for the overall efficiency of the economies of the Western countries as well as its conflicts with the trends in people's changes in values and preferences.

9. Theory and Practice in Market Economy

Economists might tell us that a free market economy will ensure people the freedom to choose which services they find the most satisfying. Accordingly, the way people live in a perfect market economy is per definition optimized. Therefore, only imperfections in the market can distort the optimal satisfaction of human needs. Economists have used similar arguments against the claim that huge potentials for technical efficiency improvements in end-use of electricity should be cost-effective. Their market argument has roughly been that, if such options were real, they would already have been made available and implemented. Nothing is perfect, however, and the real market is far from the free market theory which presently serves as an overall guideline for politics. It is a fact that these huge cost-effective potentials for more efficient end-use technologies are waiting to be utilized. And it is highly possible that equally large potentials for more efficient lifestyles are available, if we care to pick them up.

10. Market Economy in Conflict with Environmental Sustainability

A market economy is based on *individual* self interest and can therefore never by itself solve the *collective* problems, whether they are social or environmental in character. The fact mentioned above that a lot of electricity savings are very cost-effective, even from a traditional shortsighted economic view point, should not lead to the attitude that *only* cost-effective savings should be pursued. Saving a kWh of electricity is environmentally superior to almost any way to produce a kWh, so it makes good sense to pay more for saving than for producing. The conclusion is that a properly *regulated* market is necessary. To design this regulation requires foresighted and courageous politicians, as well as planners and analysts.

International liberalization of the market for electricity, for instance within EU, will probably make it more difficult to promote electricity savings. The outlook is that the burden of responsibility for saving electricity should be removed from the shoulders of electric utilities, since it is basically against their interests for the following reason. Saving electricity should be part of an Integrated Resource Planning, IRP, strategy[5]. The fundamental thesis of IRP is *to provide the electricity service, not the electricity, in an optimal way.* A company on a liberal market will have to compete on providing *electricity* in the optimal (cheapest) way, since that is what can be traded internationally, not services. Given that you cannot optimize and sub-optimize a system simultaneously, the system will not provide electricity *services* in an optimal way. Governments must necessarily take over the Integrated Resource Planning in order to achieve optimal provision of electricity services.

11. Indicators of Progress

If we agree that a better quality of life for people should be the overall aim of an economy, then the question is how to measure or just indicate whether we are making progress. In an economy where basic needs for food, shelter, clothing, and other physical goods are not well satisfied, a measure like the Gross Domestic Product, GDP, might be a reasonable indicator of the progress towards satisfying these needs. This was the situation in the 1930s, when GDP was internationally defined. Today, however, this concept has lost most of its meaning as a guideline. Economists will generally agree on that, but they still stick to GDP in lack of other indicators. Only few attempts have been made to come up with a better indicator.

Rather than trying to define one single indicator, which should encompass the values of all elements of the development, it could be more appropriate for democratic societies to establish a set of indicators for the development in for instance crime rates, income, pollution, health, etc. These indicators should then be published regularly as a basis for the democratic process of weighing the elements and making decisions. With one single indicator only, in which the elements are more or less weighed by experts and calculated by a computer, the whole democratic process will easily become obsolete as a basis for decisions, since the political value choices are already included in the computer program. Nevertheless, attempts to establish an alternative to GDP, reveal some interesting trends.

One of the attempts to define a better single indicator is what has been termed Genuine Progress Indicator, GPI, and is carried out in a project called Redefining Progress[6]. The new indicator, GPI, is actually not going very far in revising the GDP, just correcting some of the most obvious absurdities. For instance, expenses associated with crime and other social problems are counted as a cost to be deducted in the GDI, rather than being added as positive contributions as they appear in GDP. Similarly, the GPI is counting resource depletion and pollution as a cost instead of a benefit as in today's GDP. Also, the new indicator, GPI, values the possession and availability of refrigerators, furniture, cars and other so-called durable goods as positive, whereas the replacement of these goods is deducted as a cost. This is quite different from today's GDP, which includes only the replacement, and then as positive contribution. This element of the GDP is the one that encourages the buy and throw-away culture which is dominating Western societies.

In Figure 3 the development for the USA in this new indicator GPI is shown for the USA, compared to the conventional GDP. As seen, with just these rather modest adjustments of the GDP the result is striking. Since the mid 1960, the Genuine Progress Indicator has declined, reaching a level now below that in 1950. Despite this indication of how people are actually getting worse and worse off, the target of the economic policy is still the same, namely to increase the rather irrelevant GDP.

Other single indicators to replace GDP have been suggested, for instance the Index of Sustainable Economic Welfare, ISEW [7]. None of them are pretending to be a perfect measure of the development in people's general well-being. But they all seem to be superior to GDP in that respect.

It is worth noticing, that changes in people's preferences away from more income as described earlier, coincide rather well with the decline in these new indicators.

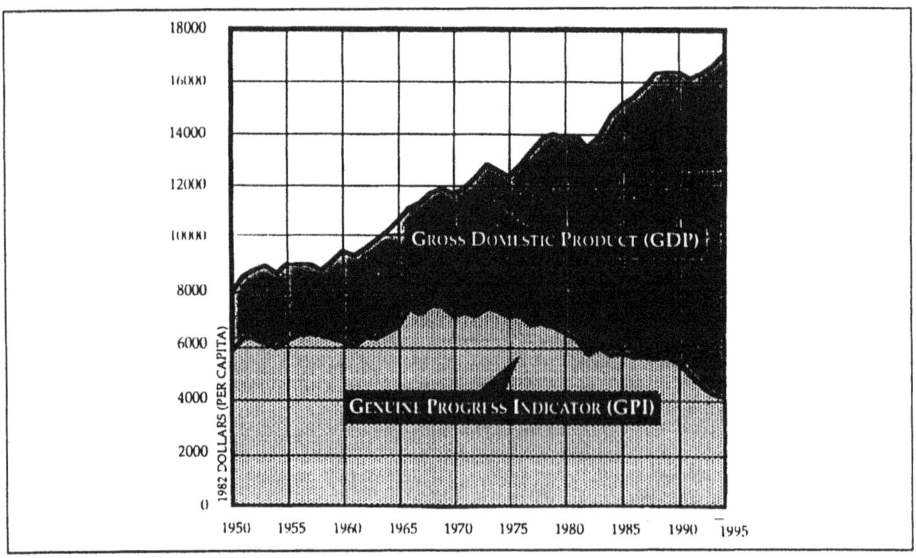

Figure 3.
The conventional Gross Domestic Product, GDP, is here compared with the new Genuine Progress Indicator, GPI. From "Redefining Progress"[6].

12. Counteracting Saturation Trends

As the saturation trends increase, they are counteracted by business and business supportive governments. Credit card and other electronic payment systems tempt consumers to buy more than they really want, because they are not forced to give the purchase an extra thought while going to the bank to fetch the cash. Shopping hours, which were introduced in Europe as a means to limit consumption and protect shop staff, are now liberated as a means to increase the consumption. Increased inequities are gradually implemented as a means to get people to work harder and buy more in a hopeless attempt to keep up. Most of all, the old sales promoting measure of advertising is an ever growing part of this counteraction. It is being allowed to interfere in more and more situations of our daily life, like theater, TV, sport, telecommunication, mail, schools, public streets, etc., with the sole overall purpose of increasing sales of various products, especially luxury items like electric appliances. Rather than opposing an economic saturation, we should, from an ecological as well as a human point of view, welcome such a trend to voluntarily slowing down a little the growth in general consumption and production.

To an increasing degree our values, and those of future generations, are being shaped through advertising and other marketing efforts, not by any democratic process, but by the business world.

12.1. LOW ELECTRICITY EUROPE AS A CASE

A study combining the technical end-use efficiencies with lifestyles efficiencies discussed in this paper, was conducted for Western Europe[8]. The methodology, assumptions and results are briefly outlined in the following.

13. Methodology for Analyzing Electricity Demand

The method used in the model for a Low Electricity Europe, is based on the following three main factors determining energy consumption.

Population is in this model simply the number of people in a country or a region, and expressed by P.

Energy Service level per capita, S, illustrates the material standard of living and hence is typically expressing the economic development.

Energy Intensity, I, of the technology used to provide the energy services is defined as the energy consumption per service provided, which is the reciprocal of efficiency.

Using indices, based on normalizing all parameters with the situation in a base year, b, energy consumption in a certain future year, y, can then be expressed as:

$$E_y = P_y * S_y * I_y .$$

Assuming, for instance that washing machines become twice as electricity-efficient as in the base year, I = 0.5, the number of laundries increases by 20%, S = 1.2, and the population does not change, then the index for electricity consumption for that particular purpose will in a future year be:

$$E_y = 1.0 * 1.2 * 0.5 = 0.6,$$

which implies a 40% lower electricity consumption for washing.

Electricity consumption in a country or region is then grouped according to two dimensions, namely 1) the end-uses like pumping, ventilation, refrigeration, lighting, etc., and 2) the economic sectors like domestic, agriculture, public service, private service, and a number of industrial sectors. This makes up a matrix, and in the Low Electricity Europe study the matrix consisted of a total of 175 elements, or uses of electricity. For each element future electricity consumption was calculated as an index by the above equation.

14. Scenario Results for Western Europe

A scenario which illustrates the combination of improved technology and development in lifestyles towards a gradual saturation was carried out in the Low Electricity Europe study.

The technological improvement is given by the electricity intensity indices, I, while the life-style was expressed by the development in the electricity service index S. The overall input of I, S, as well as population, P, summarized for Western Europe are shown in Figure 4, also showing the resulting development in electricity consumption, E.

The development of standard of living, as expressed by the electricity service S, is on the average leveling off at a standard for all countries about that in Northern Europe today. This covers a variation ranging from almost no increase in countries with a high standard today, and a 2.5 fold increase in those with a presently low standard. Technological efficiency is on average more than doubled, shown as an electricity intensity index, I, reaching down to about 45% of present. Electricity consumption can in this case almost be reduced to half of present, and what is more important, remain at that low level. Western European hydropower alone will in this scenario be able to cover close to half the supply needed, solving a lot of environmental controversies about other supply options.

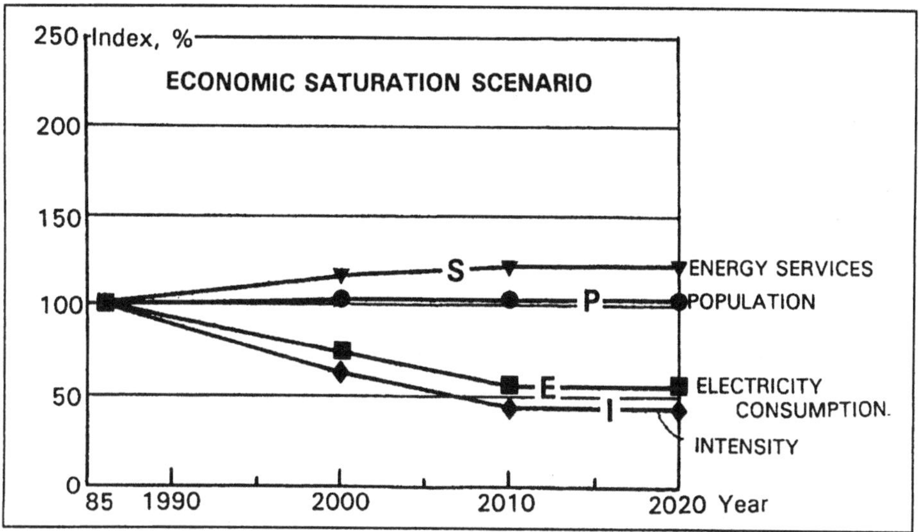

Figure 4.
The three main input parameters P, S, and I, for a scenario with a gradual saturation in the economy, together with the resulting electricity consumption in Western Europe, all in indices[8].

Figure 5 shows the same development in electricity consumption as shown in Figure 4, but now expressed in annual total consumption in TWh, split up according to the various economic sectors.

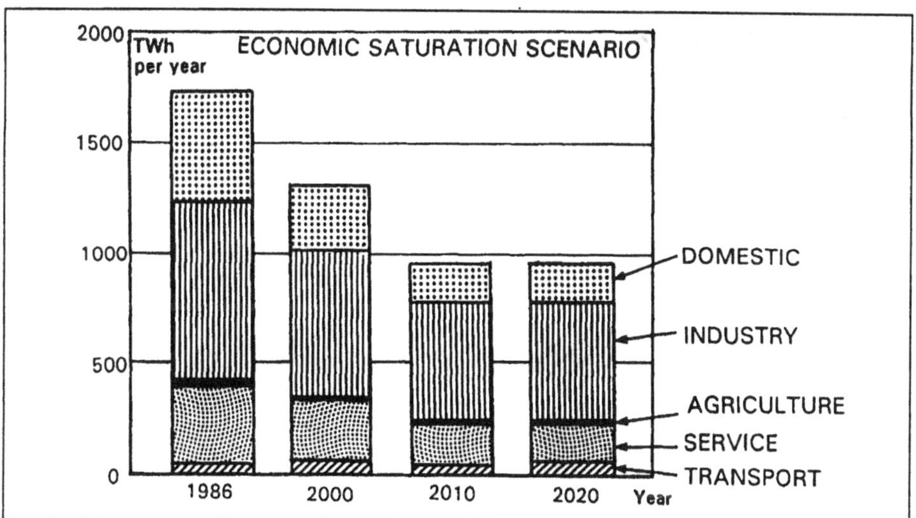

Figure 5.
Total electricity consumption in Western Europe for the scenario shown in Figure 4[8].

15. Concluding Remarks

Acknowledgment of the substantial options for using electricity more efficiently should not lead to the conclusion that this can allow us to go on for ever expanding the material production and consumption. On the other hand, there are no indications that the changes in the course of lifestyle need to hamper the quality of life, even in the most affluent countries. Often the contrary could be the case. This is also reflected in people's preferences as they are reflected in surveys as well as actual action. The changes needed seem to be more at a level of government and international organization.

Today East European countries generally consume less electricity per capita than Western Europe, and they also contribute less to the CO_2-emission. It seems possible for them to maintain this lead while improving living conditions, because changing the development in technologies as well as in economic structures in a way is easier in the East European countries, which are in a *technological* as well as an *economic* transition. It is a responsibility for the West to encourage and support such a development with sustainable use of electricity in the East as well as in the West itself.

16. References

1. Nørgård, Jørgen S.: "Sustainable Use of Electricity", in *Development with Sustainable Use of Electricity*, Kluwer Academic Publisher, Dordrecht, The Netherlands, 1998.
2. Jespersen, Jesper: "Reconciling Environment and Employment - Switching from Goods to Services", paper presented at Eco-Efficient Service Seminar, Wuppertal Institute, Germany, 18-20 September 1994.
3. Packard, V.: "The Hidden Persuaders", David McKay Compaby, Inc., New York, 1957.
4. Platz, M.: "Working Time" (in Danish), The National Institute of Social Research, Copenhagen, 1988.
5. Almeida, A.T., A.H. Rosenfeld, J. Roturier, and J. Norgard (eds.): "Integrated Electricity Resource Planning", Kluwer Academic Publisher, Dordrecht, The Netherlands, 1994.
6. Genuine Progress Indicator: "The Genuine Progress Indicator. Summary of Data and Methodology", Redefining Progress, One Kearny Street, Fourth Floor, San Fransisco, CA 94108, USA, 1995.
7. Daly, H. and J. Cobb: "For the Common Goods", Beacon Press, Boston, 1994.
8. Nørgård, J.S. and J. Viegand: "Low Electricity Europe - Sustainable Options", Report published by European Environmental Bureau, Rue de la Victoire 26 bte 12, B-1060 Brussels, Belgium, 1994.

SUBJECT INDEX

Kolibri 360 GmbH
Ferdinand-Jühlke-Straße 7
99095 Erfurt, Deutschland
produktsicherheit@kolibri360.de